高等职业教育"十三五"规划教材（网络工程课程群）

Java Web 应用开发

主　编　万　青　杨智勇
副主编　王海洋　梅青平　谢　伟

中国水利水电出版社
·北京·

内 容 提 要

本书以项目导向、任务驱动的形式，通过用户注册、用户信息管理和图书在线销售系统 3 个模块，全面、详细地介绍了开发 Java Web 应用项目所需要的各种知识与技能。相关知识内容包括软件开发环境搭建、HTML 和 CSS 基础、JSP 运行原理、页面组成和内置对象、数据库访问分层设计、Servlet 和 EL 表达式、Session 和 Cookie 的应用、JSTL 标签的应用等。作者结合多年的项目开发经验和教学实践经验，依据企业项目实施中的能力要求以及学生的认知规律，并请教大量在一线工作的软件开发工程师，完成了本书编写。

本书可作为高职高专院校和应用型本科院校计算机及网络相关专业的学生学习 Java Web 应用开发的教材和参考书。

图书在版编目（CIP）数据

Java Web 应用开发 / 万青，杨智勇主编. -- 北京：中国水利水电出版社，2017.5
高等职业教育"十三五"规划教材. 网络工程课程群
ISBN 978-7-5170-5323-1

Ⅰ. ①J… Ⅱ. ①万… ②杨… Ⅲ. ①JAVA 语言－程序设计－高等学校－教材 Ⅳ. ①TP312.8

中国版本图书馆 CIP 数据核字(2017)第 076624 号

策划编辑：祝智敏　责任编辑：李 炎　加工编辑：郭继琼　封面设计：梁 燕

书　名	高等职业教育"十三五"规划教材（网络工程课程群） Java Web 应用开发　Java Web YINGYONG KAIFA
作　者	主　编　万　青　杨智勇 副主编　王海洋　梅青平　谢　伟
出版发行	中国水利水电出版社 （北京市海淀区玉渊潭南路 1 号 D 座 100038） 网　址：www.waterpub.com.cn E-mail：mchannel@263.net（万水） 　　　　sales@waterpub.com.cn 电　话：（010）68367658（营销中心）、82562819（万水）
经　售	全国各地新华书店和相关出版物销售网点
排　版	北京万水电子信息有限公司
印　刷	三河市铭浩彩色印装有限公司
规　格	184mm×260mm　16 开本　10.5 印张　226 千字
版　次	2017 年 5 月第 1 版　2017 年 5 月第 1 次印刷
印　数	0001—3000 册
定　价	32.00 元

凡购买我社图书，如有缺页、倒页、脱页的，本社营销中心负责调换

版权所有·侵权必究

丛书编委会

主　　任：杨智勇　李建华

副主任：王璐烽　武春岭　乐明于　任德齐　邓　荣
　　　　黎红星　胡方霞

委　　员：万　青　王　敏　邓长春　冉　婧　刘　宇
　　　　　刘　均　刘海舒　刘　通　杨　坝　杨　娟
　　　　　杨　毅　吴伯柱　吴　迪　张　坤　罗元成
　　　　　罗荣志　罗　勇　罗脂刚　周　桐　单光庆
　　　　　施泽全　宣翠仙　唐礼飞　唐　宏　唐　林
　　　　　唐继勇　陶洪建　麻　灵　童　杰　曾　鹏
　　　　　谢先伟　谢雪晴

序 言

《国务院关于积极推进"互联网+"行动的指导意见》的发布标志着我国全面开启通往"互联网+"时代的大门,我国在全功能接入国际互联网20年后达到全球领先水平。目前,我国约93.5%的行政村已开通宽带,网民人数超过6.5亿,一批互联网和通信设备制造企业进入国际第一阵营。互联网在我国的发展,分别"+"出了网购、电商,"+"出了O2O(线上线下联动),也"+"出了OTT(微信等顶端业务),2015年全面进入"互联网+"时代,拉开了融合创新的序幕。纵观全球,德国通过"工业4.0战略"让制造业再升级;美国通过"产业互联网"让互联网技术的优势带动产业提升;如今在我国,信息化和工业化的深度融合越发使"互联网+"被寄予厚望。

"互联网+"时代的到来,使网络技术成为信息社会发展的推动力。社会发展日新月异,新知识、新标准层出不穷,不断挑战着学校相关专业教学的科学性,这给当前网络专业技术人才的培养提出了极大的挑战。因此,新教材的编写和新技术的更新也显得日益迫切。教育只有顺应时代的需求持续不断地进行革命性的创新,才能走向新的境界。

在这样的背景下,中国水利水电出版社和重庆工程职业技术学院、重庆电子工程职业学院、重庆城市管理职业学院、重庆工业职业技术学院、重庆信息技术职业学院、重庆工商职业学院、浙江金华职业技术学院等示范高职院校,以及中兴通讯股份有限公司、星网锐捷网络有限公司、杭州华三通信技术有限公司等网络产品和方案提供商联合,组织来自企业的专业工程师和部分院校的一线教师协同规划和开发了本系列教材。教材以网络工程实用技术为脉络,依托自企业多年积累的工程项目案例,将目前行业发展中最实用、最新的网络专业技术汇集到专业方案和课程方案中,然后编写入专业教材,再传递到教学一线,以期为各高职院校的网络专业教学提供更多的参考与借鉴。

一、整体规划全面系统　紧贴技术发展和应用要求

本系列教材的规划和内容的选择都与传统的网络专业教材有很大的区别,选编知识具有体系化、全面化的特征,能体现和代表当前最新的网络技术的发展方向。为帮助读者建立直观的网络印象,本书引入来自企业的真实网络工程项目,让读者身临其境地了解发生在真实网络工程项目中的场景,了解对应的工程施工中所需要的技术,学习关键网络技术应用对应的技术细节,对传统课程体系实施改革。真正做到以强化实际应用,全面系统培养人才,尽快适应企业工作需求为教学指导思想。

二、鼓励工程项目形式教学　知识领域和工程思想同步培养

倡导教学以工程项目的形式开展,按项目划分小组,以团队的方式组织实施;倡导各团队成员之间进行技术交流和沟通,共同解决本组工程方案的技术问题,查询相关技术资料,并撰写项目方案等工程资料。把企业的工程项目引入到课堂教学中,针对工程中所需要的实际工作技能组织教学,重组理论与实践教学内容,让学生在掌握

理论体系的同时，能熟悉网络工程实施中的实际工作技能，缩短学生未来在企业工作岗位上的适应时间。

三、同步开发教学资源　及时有效更新项目资源

为保证本系列课程在学校的有效实施，丛书编委会还专门投入了巨大的人力和物力，为本系列课程开发了相应的、专门的教学资源，以有效支撑专业教学实施过程中备课、授课、项目资源的更新和疑难问题的解决，详细内容可以访问中国水利水电出版社万水分社的网站，以获得更多的资源支持。

四、培养"互联网+"时代软技能　服务现代职教体系建设

互联网像点石成金的魔杖一般，不管"+"上什么，都会发生神奇的变化。互联网与教育的深度拥抱带来了教育技术的革新，引起了教育观念、教学方式、人才培养等方面的深刻变化。正是在这样的机遇与挑战面前，教育在尽量保持知识先进性的同时，更要注重培养人的"软技能"，如沟通能力、学习能力、执行力、团队精神和领导力等。为此，在本系列教材规划的过程中，一方面注重诠释技术，一方面融入了"工程""项目""实施"和"协作"等环节，把需要掌握的技术元素和工程软技能一并考虑进来，以期达到综合素质培养的目标。

本系列教材是出版社、院校和企业联合策划开发的成果，希望能吸收各方面的经验，集众所长，保证规划课程的科学性。配合专业改革、专业建设的开展，丛书主创人员先后组织数次研讨会进行交流、修订以保证专业建设和课程建设具有科学的指向性。来自中兴通讯股份有限公司、星网锐捷网络有限公司、杭州华三通信技术有限公司的众多专业工程师，以及产品经理罗荣志、罗脂刚、杨毅等为全书提供了技术和工程项目方案的支持，并承担全书技术资料的整理和企业工程项目的审阅工作。重庆工程职业技术学院的杨智勇、李建华，重庆工业职业技术学院的王璐烽，重庆电子工程职业学院的武春岭、唐继勇，重庆城市管理职业学院的乐明于、罗勇，重庆工商职业学院的胡方霞，重庆信息技术职业学院的曾鹏，浙江金华职业技术学院的宣翠仙等在全书成稿过程中给予了悉心指导及大力支持，在此一并表示衷心的感谢！

本系列教材的规划、编写与出版历经三年的时间，在技术、文字和应用方面历经多次的修订，但考虑到前沿技术、新增内容较多，加之作者文字水平有限，错漏之处在所难免，敬请广大读者批评指正。

丛书编委会

前言

Web 应用程序具有跨平台、跨系统的特点，不论是在 PC 上、手机上还是平板电脑上，不论是用 Windows 系统、Linux 系统还是 Mac 系统，只要安装了浏览器，就可以通过访问 Web 服务器来运行 Web 应用。

基于 Java 平台的 Web 应用开发，需要一些基础知识作为铺垫，包括 Java 语言基础、网页语言基础（HTML 和 CSS）和数据库基础。本书以项目导向、任务驱动的形式，循序渐进地完成一些具体的工作任务，并在此基础上进行改进，获得软件设计思想和理念上的提升。在设计和实现过程中尽量简明、高效，力求接近最佳实践，在达到能用 JSP 完成 Web 应用项目的基础上，为以后进一步学习 Java EE 框架作好准备。

1. 本书内容

项目 1　开发环境安装与配置

掌握 Java Web 应用程序开发环境的安装与配置。整个开发环境包括 JDK、Eclipse 和 Tomcat 这三个组成部分。搭建好开发环境后，在此基础上创建并运行一个简单的 Java Web 应用程序。

项目 2　实现用户注册功能

掌握网页表单标签的用法，以及通过表单提交数据后的接收、处理和响应。具体任务：设计用户注册表单，接收表单提交的数据，验证提交的数据是否合法，处理表单提交的数据并响应输出。

项目 3　改进用户注册功能

掌握 JDBC 的封装，数据库操作的分层设计，用 Servlet 接收和处理数据，用 EL 表达式显示数据。具体任务：用 Servlet 接收表单提交的数据，验证提交数据的合法性，用封装后的 DBHelper 类在数据库中检验和保存用户数据，用 EL 表达式显示输出结果。

项目 4　实现用户管理功能

掌握 Cookie 和 Session 这两种会话跟踪机制，用 EL 表达式和 JSTL 标签实现 JSP 中的数据显示。具体任务：采用 MVC 模式，设计管理员登录模块和用户列举、添加、删除、修改模块。

项目 5　实现网上书店

以网上书店应用系统为例，了解 Web 应用系统的用户角色和功能模块划分方法，掌握 Web 应用系统的需求分析、系统设计、代码编写和测试等设计流程。完成的功能模块包括：用户注册、登录、浏览、搜索图书，选购图书，提交订单。

2. 本书特点

（1）以项目为导向，通过具体、简明的工作任务来驱动整个学习过程。本书的每个项目中都先提出要完成的任务，再学习相关知识，分步骤完成，目标明确，可操作性强。

（2）由浅入深，循序渐进。本书的任务采用阶梯式编排，每个任务完成后技术水平和思维都将上升一个台阶，技术路线清晰，有利于知识和技能的巩固。

（3）承前启后，顺应行业发展趋势。本书的内容是前导课程"Java 程序设计""网页设计基础""数据库基础"的综合应用，也是后续课程"Java EE 项目开发"的铺垫。在内容编排上，将数据库分层设计、MVC 架构作为重点讲解，不仅衔接前导课程，还为以后学习 SSH（或 SSM）等项目架构做好充分准备。

（4）学以致用，注重能力拓展。每个任务都以"任务分析—相关知识—任务实施—实践训练"为线索进行编写，重视实践能力、自学能力、拓展能力的培养。

（5）提供丰富的教学资源。本书提供教学用的 PPT 课件、课程案例、项目代码等资源下载，方便教师授课和学生学习。

3. 读者定位

本书可作为高职高专院校和应用型本科院校计算机及网络相关专业的学生学习 Java Web 应用开发的教材和参考书。

4. 作者团队

本书的作者团队由教学经验和工程项目经验丰富的一线骨干教师组成，由万青、杨智勇担任主编，王海洋、梅青平、谢伟担任副主编。其中项目 1、项目 5 由万青编写，项目 2 由杨智勇编写，项目 3、项目 4 由王海洋编写，由万青审稿。另外，参与本书部分编写工作的还有：梅青平、谢伟、段萍、邱雷、郑小蓉等。本书在编写过程中，得到了重庆工程职业技术学院吴再生副校长、重庆工程职业技术学院信息工程学院李建华书记的关心和支持，在此表示感谢。

由于编者水平有限，疏漏之处在所难免，敬请读者批评指正。

编者

2017 年 3 月

目录

序言
前言

项目 1　开发环境安装与配置 .. 001
单元介绍 001
学习目标 001

任务 1.1　安装、配置 JDK 和 Eclipse 002
【任务分析】 002
【相关知识】 002
　1.1.1　JDK 简介 002
　1.1.2　Eclipse 简介 003
【任务实施】 003
【实践训练】 007

任务 1.2　安装、配置 Tomcat 008
【任务分析】 008
【相关知识】 008
【任务实施】 008
【实践训练】 014

任务 1.3　创建第一个 Web 应用程序项目 015
【任务分析】 015
【相关知识】 015
　1.3.1　Web 应用程序存放位置与结构 015
　1.3.2　Web 项目的组织结构 015
【任务实施】 016
【实践训练】 017

拓展训练 018
同步训练 018

项目 2　实现用户注册功能 020
单元介绍 020
学习目标 020

任务 2.1　设计注册信息输入页面 021
【任务分析】 021
【相关知识】 021
　2.1.1　HTTP 021
　2.1.2　HTML 022
　2.1.3　表单标签 023
　2.1.4　CSS 定位与 DIV 布局 025
【任务实施】 032
【实践训练】 034

任务 2.2　处理表单提交的数据 034
【任务分析】 034
【相关知识】 034
　2.2.1　JSP 的运行原理 034
　2.2.2　JSP 页面的组成部分 035
　2.2.3　JSP 内置对象 038
　2.2.4　表单提交数据的接收 041
　2.2.5　验证输入信息 042
　2.2.6　响应输出到浏览器 044
【任务实施】 045
【实践训练】 047

拓展训练 048
同步训练 048

项目 3　改进用户注册功能 049
单元介绍 049
学习目标 049

任务 3.1　验证和保存用户信息 050
【任务分析】 050
【相关知识】 050
　3.1.1　JDBC 的基本用法 050
　3.1.2　JDBC 的封装 057
　3.1.3　数据库操作分层设计 061

【任务实施】................................. 065
　　【实践训练】................................. 068

任务 3.2　用 Servlet 接收和处理数据..................068
　　【任务分析】................................. 068
　　【相关知识】................................. 068
　　　3.2.1　Servlet 基础................... 068
　　　3.2.2　EL 表达式....................... 077
　　【任务实施】................................. 081
　　【实践训练】................................. 086

拓展训练... 086

同步训练... 086

项目 4　实现用户管理功能......087

单元介绍... 087

学习目标... 087

任务 4.1　管理员登录........................088
　　【任务分析】................................. 088
　　【相关知识】................................. 088
　　【任务实施】................................. 097
　　【实践训练】................................. 104

任务 4.2　用户管理............................ 104
　　【任务分析】................................. 104
　　【相关知识】................................. 104
　　【任务实施】................................. 118
　　【实践训练】................................. 130

拓展训练... 130

同步训练... 130

项目 5　实现网上书店..........131

单元介绍... 131

学习目标... 131

任务 5.1　用户登录和图书展示模块设计..................132
　　【任务分析】................................. 132
　　【相关知识】................................. 132
　　　5.1.1　网上书店系统需求分析..... 132
　　　5.1.2　功能模块设计................... 132
　　【任务实施】................................. 133
　　【实践训练】................................. 145

任务 5.2　购物和订单生成模块设计..................145
　　【任务分析】................................. 145
　　【相关知识】................................. 145
　　　5.2.1　网上商城购物车............... 145
　　　5.2.2　购物车的数据存储方式..... 146
　　【任务实施】................................. 146
　　【实践训练】................................. 156

拓展训练... 156

同步训练... 156

参考文献..................................158

项目 1
开发环境安装与配置

单元介绍

通过本项目,掌握 Java Web 应用程序开发环境的安装与配置。整个开发环境包括 JDK、Eclipse 和 Tomcat 这三个组成部分。搭建好开发环境后,在此基础上创建并运行一个简单的 Java Web 应用程序。

本项目包括以下任务:
- 安装、配置 JDK 和 Eclipse
- 安装、配置 Tomcat
- 编写一个 Web 应用程序

学习目标

【知识目标】
　　了解 JDK、Eclipse、Tomcat 的概念
　　掌握 JDK、Eclipse、Tomcat 的下载、安装方法
　　掌握 Eclipse 的配置和使用方法
　　掌握 Tomcat 的配置方法
　　了解 Java Web 应用程序的基本结构

【能力目标】
　　能下载、安装 JDK
　　能下载、安装、配置和使用 Eclipse
　　能下载、安装、配置 Tomcat
　　能创建并运行一个简单的 Java Web 应用程序

任务 1.1　安装、配置 JDK 和 Eclipse

【任务分析】

在进行 Java Web 应用程序开发之前，要搭建好开发平台，本节任务主要包括 JDK 和 Eclipse 的下载和安装。其中 JDK 是 Java 开发工具集，为 Java 程序的编译、执行提供底层支持，Eclipse 是一个可视化的软件项目开发工具，具有组织和管理项目文件、代码智能提示、自动完成，项目编译部署等功能，Eclipse 的功能可以通过插件扩展。

【相关知识】

1.1.1　JDK 简介

JDK 是 Java Development Kit 的简称，是 Java 语言的开发工具包，它包含了 Java 的运行环境（JRE）、工具和基础类库（rt.jar）。

最主流的 JDK 是 Sun 公司发布的 JDK，从 Sun 的 JDK5.0 开始，提供了泛型等实用的功能，其版本名称也不再延续以前的 1.1、1.2、1.3、1.4，而是变成了 5.0、6.0。2010 年 3 月，Oracle 收购 Sun Microsystems，目前 JDK 的最新版本是 8.0。

JDK 中常用的组件有：

javac——编译器，将源程序转成字节码。

jar——打包工具，将相关的类文件打包成一个文件。

javadoc——文档生成器，从源码注释中提取文档。

jdb——调试工具。

java——运行编译后的 Java 程序（.class 文件）。

appletviewer——小程序浏览器。

另外，JDK 中还包含了一些基础的包：

java.lang——包含一些 Java 语言的核心类，如 String、Math、Integer、System 和 Thread，在 java.lang 包中还有一个子包 java.lang.reflect，用于实现 Java 类的反射机制。

java.awt——包含了构成抽象窗口工具集（abstract window toolkits）的多个类，这些类用来构建应用程序的图形用户界面（GUI）。

javax.swing——Swing 是在 AWT 的基础上构建的一套新的图形界面系统，它提供了 AWT 的所有功能，并且用纯粹的 Java 代码对功能进行了大幅度的扩充。

java.applet——包含 Applet 运行所需的一些类。

java.net——包含执行与网络相关操作的类。

java.io——包含提供输入/输出功能的类。

java.util——包含一些实用工具类。

java.sql——包含数据库操作类。

1.1.2　Eclipse 简介

Eclipse 是一个开放源代码的、基于 Java 的应用开发平台，该平台本身是一个具有可扩展性的框架，可以通过插件进行扩展。Eclipse 最初是由 IBM 公司开发的集成开发环境，2001 年 11 月贡献给开源社区，现在由非营利软件供应商联盟 Eclipse 基金会管理。2007 年 6 月，稳定版 3.3 发布；2008 年 6 月发布代号 Ganymede 的 3.4 版；2009 年 6 月发布代号 Galileo 的 3.5 版；2010 年 6 月发布代号 Helios 的 3.6 版；2011 年 6 月发布代号 Indigo 的 3.7 版；2012 年 6 月发布代号 Juno 的 4.2 版；2013 年 6 月发布代号 Kepler 的 4.3 版；2014 年 6 月发布代号 Luna 的 4.4 版；2015 年 6 月发布代号 Mars 的 4.5 版。

【任务实施】

1. 下载 JDK

从 http://www.java.com/ 可以下载 JDK 的最新版本。官方主页如图 1.1 所示。

图 1.1　JDK 下载页面

目前可以在官方网站下载 JDK8，如果要下载旧版本的 JDK，如 JDK7，则点击"所有 Java 下载"链接，下载旧版本的 JDK，或通过搜索引擎搜索关键字"JDK7"，根据搜索结果，进入相关页面下载。下载时注意根据操作系统类型选择合适的安装程序。

2. 安装 JDK

下载后运行安装程序，出现图 1.2 所示的安装向导。

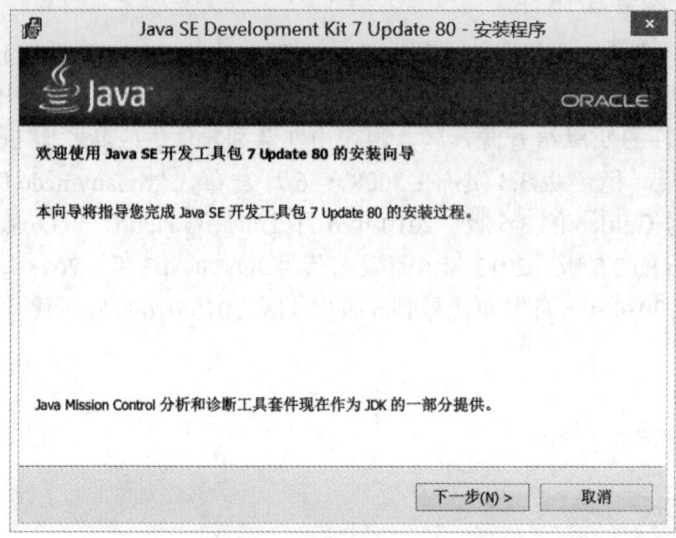

图 1.2　JDK 安装向导

点击"下一步"，选择要安装的组件和安装位置，点击"下一步"完成安装，如图 1.3 所示。

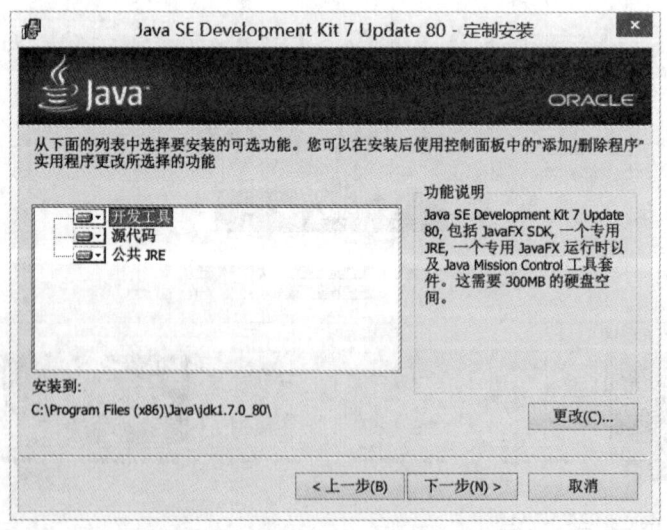

图 1.3　安装项目和安装目录设置

3. 下载 Eclipse

在 Eclipse 官方网站 http://www.eclipse.org/downloads/ 的下载页面可以下载安装程

序。开发 Java Web 应用程序应下载"Eclipse IDE for Java EE Developers",如图 1.4 所示。进入子页面后进一步选择软件版本和对应的操作系统,这里选择最新版 4.5(代号 Mars)。

图 1.4　下载 Eclipse IDE for Java EE Developers

4. 安装和启动 Eclipse

Eclipse 是以压缩包形式发布的,解压后运行 eclipse.exe 即可。Eclipse 启动后会弹出"工作区选择"对话框,通过对话框可以设置工作区目录,该目录为 Eclipse 项目文件的存放目录,如图 1.5 所示。

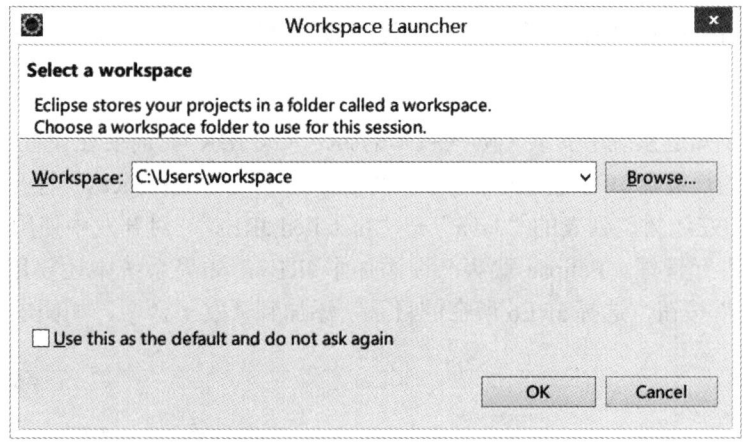

图 1.5　设置 Eclipse 工作区目录

此时可以点击"Browse"按钮选择一个目录,或直接输入路径,再点击"OK"按钮进入 Eclispe 的主界面。

5. Eclipse 启动参数配置

Eclipse 的启动由 %ECLIPSE_HOME%/eclipse.ini 控制,如果系统中没有定义环境变量 ECLIPSE_HOME,则 Eclipse 启动程序会默认读取安装目录下的 eclipse.ini。

eclipse.ini 是一个文本文件,其内容相当于在 Eclipse 启动时添加到 eclipse.exe 之后的命令行参数，该文件可以用记事本打开进行编辑。

常见的 eclipse.ini 内容如下：

```
-showsplash
org.eclipse.platform
--launcher.defaultAction
openFile
--launcher.appendVmargs
-vm
E:\jdk1.7\bin\javaw.exe
-vmargs
-Dosgi.requiredJavaVersion=1.7
-Xms256m
-Xmx1024m
```

文件中的 showsplash 表示启动时显示 logo 的闪屏设置，launcher.defaultAction 表示启动后的默认动作，-vm 表示运行 eclipse 使用的 Java 虚拟机路径，-vmargs 表示 Java 虚拟机的参数，这里要求最低 JDK 版本是 1.7，虚拟机占用的最小内存是 256MB，最大内存是 1024MB。

如果设置了 -vm 参数，可以不用在环境变量中设置 JAVA_HOME，即可正常启动 Eclipse。-vm 参数设置要求如下：

① -vm 选项和它的参数值（路径）必须在单独的一行；

② 其值必须准确地指向 Java 可执行文件，而不只是 JDK 的根目录；

③ 所有在 -vmargs 之后的参数将会被传输给 JVM，对 Eclipse 设置的参数必须写在 -vmargs 之前（就如同在命令行中加上这些参数一样）。

6. 在 Eclipse 中配置 JRE

Eclipse 不知道系统中安装了哪些版本的 JRE（或 JDK），需要在设置窗口中手动指定。Eclipse 的设置窗口通过点击菜单项"Window"—"Preferences"打开。

选择窗口左边树型列表的"Java"—"Installed JREs"，对开发中使用的 JRE（Java 运行环境）进行管理。Eclipse 默认已经添加了 JRE7，如果系统中还要用到 JRE6，则点击"Add..."按钮，选择 JRE6 所在的目录，添加到开发平台中，如图 1.6 所示。

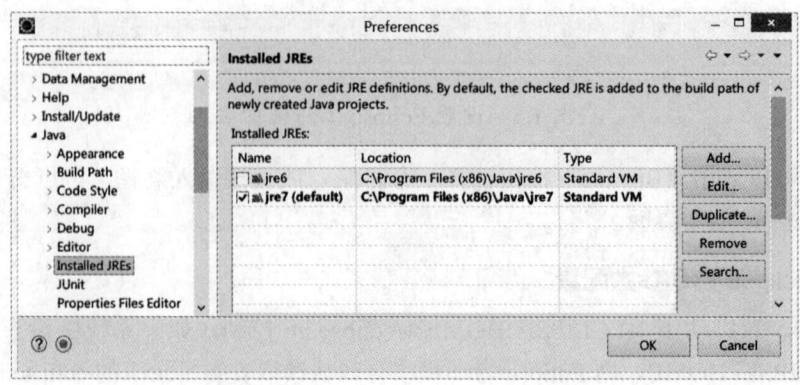

图 1.6　在 Eclipse 中设置 JRE

JRE 是支撑 Java 程序运行的基础，新建项目时，默认的 JRE 会被添加到项目的"Build Path"中，在项目编译、运行时调用。

7. 编写 Java 程序测试开发环境

在 Eclipse 中新建一个 Java 项目，起名为 Test，选择 JDK1.7（JDK7.0）作为项目编译运行环境，如图 1.7 所示。

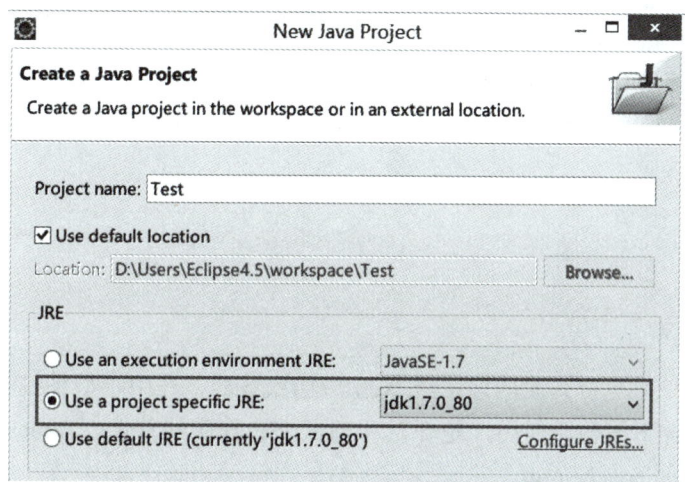

图 1.7　选择 JDK7.0 作为项目编译运行环境

在项目源码中添加 TestEnv 类，编写主函数如下：
检测 Java 运行环境参数

```java
import java.util.Properties;
public class TestEnv {
  public static void main(String[] args) {
    Properties props = System.getProperties();
    System.out.println("Java 运行环境版本："+ props.getProperty("java.version"));
    System.out.println("Java 运行环境供应商："+ props.getProperty("java.vendor"));
    System.out.println("Java 供应商 URL："+ props.getProperty("java.vendor.url"));
    System.out.println("Java 安装路径："+ props.getProperty("java.home"));
    System.out.println("Java 的类路径："+ props.getProperty("java.class.path"));
    System.out.println(" 加载库时搜索的路径列表："+ props.getProperty("java.library.path"));
    System.out.println(" 默认的临时文件路径："+ props.getProperty("java.io.tmpdir"));
  }
}
```

【实践训练】

下载并安装 JDK1.7 和 Eclipse IDE for Java EE Developers 4.5，在 Eclipse 中设置默认的 JRE，并创建 Java 项目，编写程序检测 Java 运行环境参数。

任务 1.2　安装、配置 Tomcat

【任务分析】

Java Web 应用程序在运行时不仅要依赖 JRE，还要依赖 Tomcat 容器。本节的任务是下载、安装和配置 Tomcat，并在 Eclipse 中进行设置，让 Eclipse 和 Tomcat 协同工作。

【相关知识】

1.2.1　Tomcat 简介

Tomcat 是一个免费开源的轻量级 Web 应用服务器，在中小型系统和并发访问用户不是很多的场合下被普遍使用，是开发和调试 JSP 程序的首选。Tomcat 的官方网站地址是 http://tomcat.apache.org/。

【任务实施】

1. 下载和安装 Tomcat

打开 Tomcat 的官方网站 http://tomcat.apache.org/，在主页的左边列出了各个版本的下载链接，目前最高版本是 9.0，如图 1.8 所示。

图 1.8　Tomcat 官方网站首页

选择版本后，还要进一步根据操作系统选择安装包，Tomcat 分为安装版和免安装版，安装版可以以 Windows 服务方式在后台运行，免安装版以 Windows 应用程序方式运行。在开发过程中，一般用 Eclipse 来管理 Tomcat，下载免安装版，解压即可。

2. 在 Eclipse 中配置 Tomcat

点击菜单项"Window"—"Preferences"打开 Eclipse 的设置窗口，选择窗口左边树型列表中的"Server"—"Runtime Environments"，对 Web 服务器进行管理。点击"Add..."按钮，弹出图 1.9 所示的窗口，选择对应的 Tomcat 版本，点击"Next>"，在图 1.10 所示的窗口中选择 Tomcat 的路径和 JRE 版本。

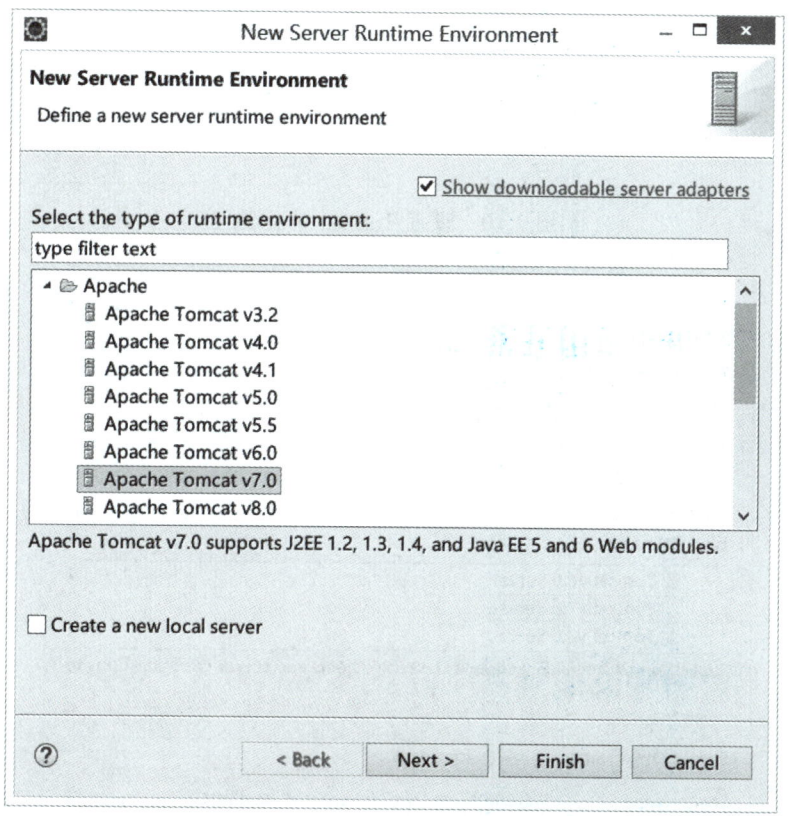

图 1.9　选择 Tomcat 服务器版本

3. 在 Eclipse 中新建 Web 服务器

Eclipse 在"Servers"窗口中列出所有的服务器，初次运行 Eclipse 时，服务器列表为空，此时点击里面的链接"Click this link to create a new server..."可以新建服务器。

点击链接后，在弹出的"New Server"窗口中选择服务器类型（如图 1.11 所示），再点"Finish"按钮完成服务器创建。在此界面中，可以点击"Add..."对选中的服务器类型进行设置。

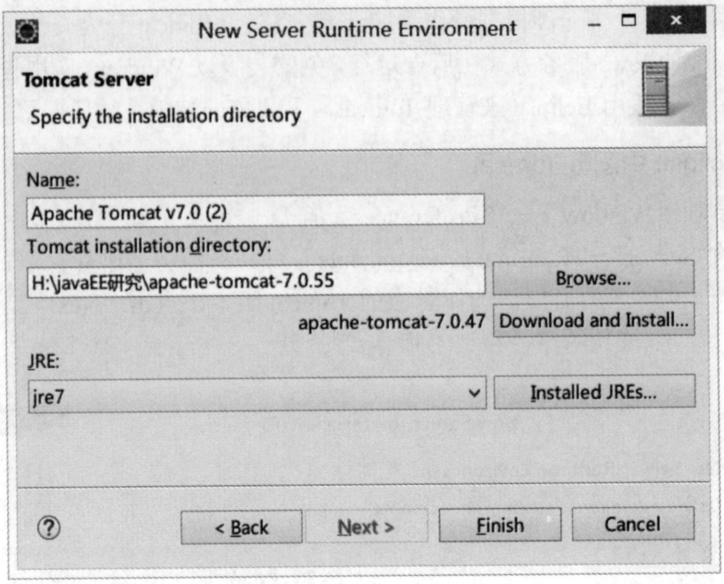

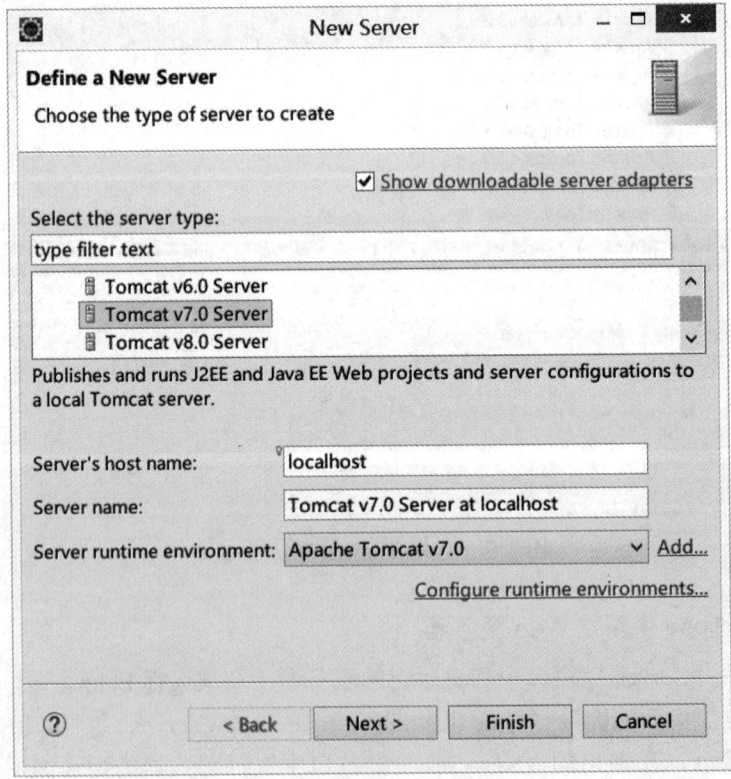

图 1.10 设置 Tomcat 服务器

图 1.11 创建服务器窗口

创建服务器后,在"Servers"窗口的服务器列表中,双击刚才添加的"Tomcat

v7.0 Server at localhost"，在 "Overview" 窗口中，将 "Server Locations" 设置为 "Use Tomcat installation"，如图 1.12 所示。

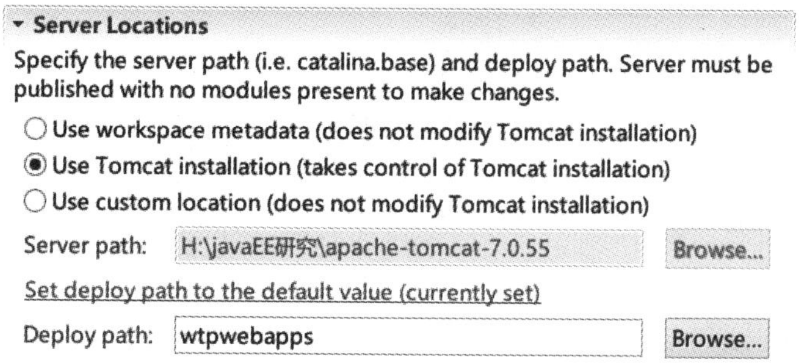

图 1.12 设置服务器位置

设置完毕注意保存，然后在 "Servers" 窗口中选中 "Tomcat v7.0 Server at localhost"，点击鼠标右键，在弹出的菜单中选择 "Start" 启动服务器。启动后服务器状态变成 "Started"，如图 1.13 所示。

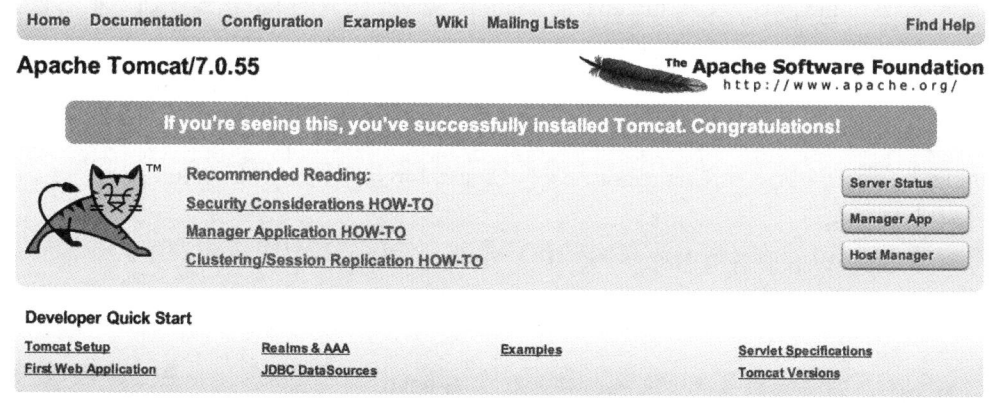

图 1.13 已经启动的服务器

服务器正确启动后，打开浏览器，输入 "http://localhost:8080/"，如果能打开如图 1.14 所示的 Tomcat 主页，说明服务器运行正常。

图 1.14 Tomcat 主页

4. 配置 Tomcat 服务器

Tomcat 服务器的主要配置文件是 server.xml 和 web.xml，在 server.xml 中配置 Web 服务器相关参数，在 web.xml 中配置 Web 应用程序参数。

（1）server.xml

Tomcat 服务器的结构如图 1.15 所示。

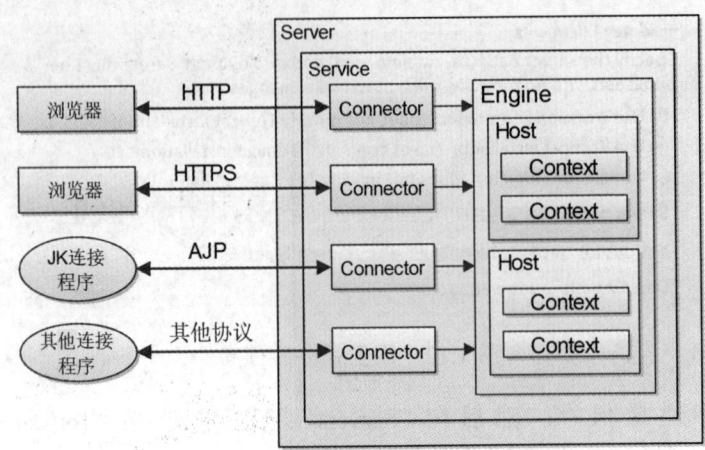

图 1.15　Tomcat 服务器的组织结构

从图中可以看出，Tomcat 服务器主要由 Server、Service、Connector、Engine、Host、Context 这些元素构成，各元素的属性可以在 server.xml 中配置。

① Server 元素：代表整个容器，是 Tomcat 实例的顶层元素，对应 org.apache.catalina.Server 接口。

主要属性：
- port，监听 shutdown（终止服务）命令的端口，默认为 8005。
- shutdown 属性，需要终止服务时，向 port 指定端口发送的命令字符串，默认为"SHUTDOWN"。
- className 属性，实现 org.apache.catalina.Server 接口的类，该属性通常不用设置，默认值为 org.apache.catalina.core.StandardServer。

② Service 元素：表示一个具体的 Web 服务，对应 org.apache.catalina.Service 接口，org.apache.catalina.core.StandardService 类是 Service 接口的标准实现。它包含一个 Engine，以及一个或多个 Connector，这些 Connector 共享同一个 Engine。

主要属性：
- name，Web 服务的名称。

③ Connector 元素：表示客户端和服务之间的连接。

主要属性：
- port，监听客户端请求的端口号，默认为 8080。
- connectionTimeout，连接超时时间，以毫秒为单位。
- protocol，通信协议。
- redirectPort 属性，在处理 HTTP 请求时，收到一个 SSL 传输请求后重定向的端口号。

④ Engine 元素：Service 中的请求处理机，接收并处理来自 Connector 的请求，每

个 Service 只能有一个 Engine。

主要属性：
- defaultHost，指定处理请求的默认主机名，至少与其中一个 Host 元素的 name 属性值一致。

⑤ Host 元素：表示一个虚拟主机，一个 Engine 可以包含多个 Host。

主要属性：
- name，主机名（即访问该主机的域名），默认是 localhost、127.0.0.1 及本地 IP。
- appBase，存放应用程序的目录，可以为相对路径或绝对路径。
- autoDeploy，自动部署项目，不用重启 Tomcat。

⑥ Context 元素：表示一个 Web 应用程序，一个 Host 可以包含多个 Context。

主要属性：
- docBase，应用程序的路径或者是 WAR 文件存放的路径。
- path，表示该 Web 应用程序 URL 的前缀，一个完整的 Web 应用程序路径是 http:// 主机名 : 端口号 / 应用程序名 / 目录（文件）名，例如 http://localhost:8080/path/index.html

默认情况下，Host 包含一个 path 为空字符串的 Context，所以省略掉应用程序名也可以访问到 ROOT 目录下的网页，如 "http://localhost:8080/index.html"。如果要修改该默认 Context 的路径，可以在 Host 中添加结点：

```
<Host appBase="webapps" autoDeploy="true" name="localhost" unpackWARs="true">
<Context docBase="" path=" 指定路径 " reloadable="true"/>
</Host>
```

- reloadable，如果为 true，则 Tomcat 会自动检测应用程序的 /WEB-INF/lib 和 /WEB-INF/classes 目录的变化，自动装载新的应用程序，可以在不重启 Tomcat 的情况下更新应用程序。

（2）web.xml

在 Tomcat 目录下，web.xml 存在于两处地方，第一处是 conf/web.xml，其设定会影响所有 Web 应用程序，第二处是 WEB-INF/web.xml，其设定只影响该应用程序本身。

①默认主页（欢迎）文件设置

```
<welcome-file-list>
  <welcome-file>index.html</welcome-file>
  <welcome-file>index.htm</welcome-file>
  <welcome-file>index.jsp</welcome-file>
</welcome-file-list>
```

上面的配置将 index.html、index.htm、index.jsp 设为候选主页，排在前面的优先级别高。

②报错页面设置

```
<error-page>
  <error-code>404</error-code>
  <location>/FileNotFound.jsp</location>
</error-page>
```

```xml
<error-page>
    <exception-type>java.lang.NullPointerException</exception-type>
    <location>/Null.jsp</location>
</error-page>
```

第一个 error-page 结点含义是：当出现 404 错误（文件未找到）时，会跳转到网站根目录下的 FileNotFound.jsp 页面。第二个 error-page 结点含义是：如果网站程序执行过程中产生 NullPointerException，则会跳转到根目录下的 Null.jsp。

③会话超时的设置

```xml
<session-config>
<session-timeout>30</session-timeout>
</session-config>
```

设置 Session 的过期时间为 30 分钟。

④过滤器设置

```xml
<filter>
    <filter-name>MyFilter</filter-name>
    <filter-class>MyProject.MyFilter</filter-class>
</filter>
<filter-mapping>
    <filter-name>MyFilter</filter-name>
    <url-pattern>/*</url-pattern>
</filter-mapping>
```

定义一个页面过滤器，拦截所有请求的页面，并由 MyProject.MyFilter 类进行预处理。

【实践训练】

1. 在 Eclipse 中设置并启动 Tomcat，然后打开浏览器，输入 Tomcat 默认网址（http://localhost:8080）检测网站是否正常运行。

2. 在 server.xml 中将 Tomcat 的监听端口号改成 9000，在 Eclipse 中重新启动 Tomcat，输入新的网址（http://localhost:9000），测试网站是否正常运行，测试完毕后将端口改回 8080。

3. 配置一个虚拟主机，主机名叫 myhost，通过域名 myhost.com 访问，根目录是 d:\myhost，步骤如下：

（1）新建 d:\myhost 目录，并在里面创建 ROOT 目录，编写网页 index.html 存放在 ROOT 目录下；

（2）在 server.xml 文件的 Engine 结点下增加一个 Host 子结点：

```xml
<Host appBase="d:/myhost" autoDeploy="true" name="myhost" unpackWARs="true">
    <Alias>myhost.com</Alias>
</Host>
```

（3）修改 Windows 系统目录"C:\Windows\System32\drivers\etc"下的 hosts 文件，增加两行内容：

127.0.0.1　myhost
127.0.0.1　myhost.com

（4）在 Eclipse 中重新启动 Tomcat，打开浏览器，输入"http://myhost:8080/index.html"，或用别名"http://myhost.com:8080/index.html"访问虚拟主机中的网页。

4．在虚拟主机中增加一个名为 app1 的 Web 应用程序。

（1）在 d:\myhost\ROOT 中创建目录 app1，编写网页 index.html 存放在 app1 中；

（2）在上面的 Host 结点中增加 Context 结点，内容如下：

```
<Host appBase="d:/myhost" autoDeploy="true" name="myhost" unpackWARs="true">
    <Alias>myhost.com</Alias>
    <Context path="/app1" docBase="ROOT/app1" reloadable="true"/>
</Host>
```

（3）在 Eclipse 中重新启动 Tomcat，打开浏览器，输入"http://myhost:8080/app1/index.html"，或用别名"http://myhost.com:8080/app1/index.html"访问虚拟主机中 Web 应用 app1 下的网页。

任务 1.3　创建第一个 Web 应用程序项目

【任务分析】

在 Eclipse 中建立一个 Java Web 应用程序项目，部署到 Tomcat 服务器中运行，主页为 index.jsp。当客户端浏览器通过指定的网址访问该 Web 应用程序时，在浏览器中显示服务器的当前日期和时间。

【相关知识】

1.3.1　Web 应用程序存放位置与结构

一个 Web 应用程序对应 server.xml 配置文件中的一个 Context 结点，其存放文件的根目录位置与 Context 结点的 docBase 属性对应。如果 Host 结点中没有设置 Context 子结点，则默认的根目录就是该 appBase 指定目录下的 ROOT 子目录。

Web 应用程序根目录下有一个特殊的文件夹 WEB-INF，其中包含当前 Web 应用程序的配置文件 web.xml，仅对当前 Web 应用生效。另外在 WEB-INF 中还包含两个子目录，分别是 classes 和 lib，其中 classes 中存放 Java 类编译后生成的 .class 文件，lib 中存放 jar 文件。

1.3.2　Web 项目的组织结构

Eclipse 中的 Web 项目组织结构如图 1.16 所示。其中 \src 目录中存放 Java 源代码，\WebContent 目录中存放 jsp 文件，配置文件 web.xml 存放在目录 \WebContent\WEB-INF 中，依赖的第三方 jar 文件（如数据库驱动）存放在 \WebContent\WEB-INF\lib 中。

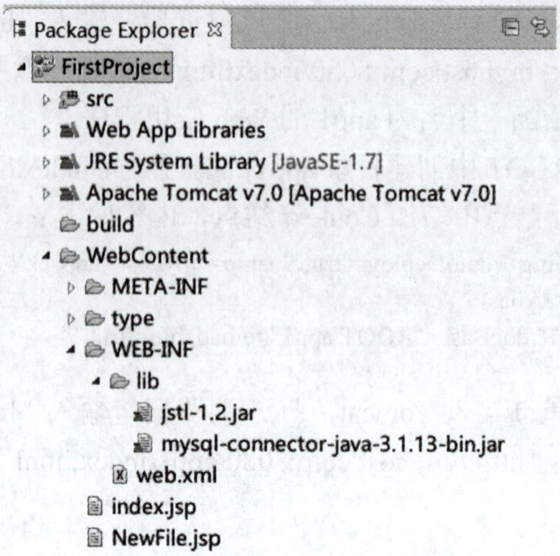

图1.16　Web项目的组织结构

【任务实施】

1. 修改Eclipse的配置文件，解决中文乱码问题

在启动Eclipse前，用写字板打开Eclipse安装目录下的配置文件eclipse.ini，在"-vmargs"栏目下增加一行，内容为"-Dfile.encoding=utf-8"，然后启动Eclipse。这项设置保证新建的JSP页面编码用UTF-8，支持中文内容。修改完毕后保存，再重新启动Eclipse。

2. 新建项目

Eclipse通过项目来管理一个Web应用程序，在编写程序之前要先新建项目。

选择菜单"File"—"New"—"Dynamic Web Project"（如果没有列出，在"Other…"中查找），在弹出的窗口中填写"Project name"为"FirstProject"，连续点击"Next"两次，在"Web Module"选项卡中勾选"Generate web.xml deployment descriptor"复选框，让项目中自动添加web.xml文件（如图1.17所示），最后点击"Finish"按钮创建新项目。

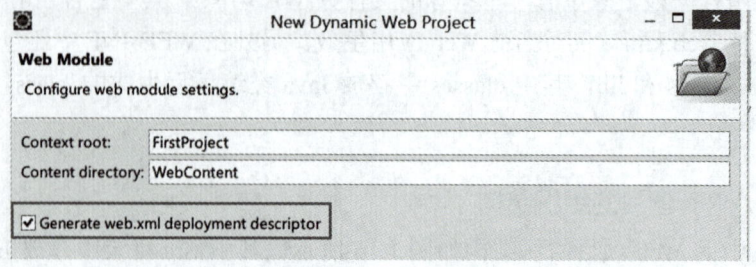

图1.17　在新建Web项目中自动添加web.xml文件

3. 新建 JSP 页面

在"Project Explorer"窗口中找到刚才创建的"FirstProject"项目，在项目的 WebContent 文件夹上点击鼠标右键，在弹出的菜单中选择"New"—"JSP File"，然后在对话框中将文件名改为"index.jsp"，点击"Finish"按钮完成 JSP 页面创建。

4. 在 JSP 页面中编写代码

```jsp
<%@ page language="java" contentType="text/html; charset=UTF-8" pageEncoding="UTF-8"%>
<%@ page import="java.util.*,java.text.*"%>
<!DOCTYPE html PUBLIC "-//W3C//DTD HTML 4.01 Transitional//EN" "http://www.w3.org/TR/html4/loose.dtd">
<html>
<head>
<meta http-equiv="Content-Type" content="text/html; charset=UTF-8">
<title> 显示当前日期时间 </title>
</head>
<body>
<%
// 设置显示格式
SimpleDateFormat df
=new SimpleDateFormat("yyyy 年 MM 月 dd 日 HH 时 mm 分 ss 秒 ");
// 获取当前日期时间
Date d=new Date();
// 输出格式化后的结果
out.println(df.format(d));
%>
</body>
</html>
```

5. 运行项目

点击工具栏上的"Run"按钮（快捷键 Ctrl+F11），在弹出的窗口中选择"Run On Server"，并在后续窗口中选择"Tomcat v7.0 Server at localhost"作为运行 Web 应用程序的服务器。如果希望后面运行的时候也用该服务器，不再弹出窗口，可以勾选"Always use this server when running this project"选项，如图 1.18 所示。

运行后，在 Eclipse 的"Console"窗口中会显示出服务器的启动过程，并打开浏览器显示运行结果。如果 Eclipse 自带的浏览器没有自动打开，可以在外部打开浏览器，输入网址"http://localhost:8080/FirstProject/index.jsp"查看运行结果。

【实践训练】

创建一个 Web 应用程序项目，在 JSP 页面中打印乘法九九表。

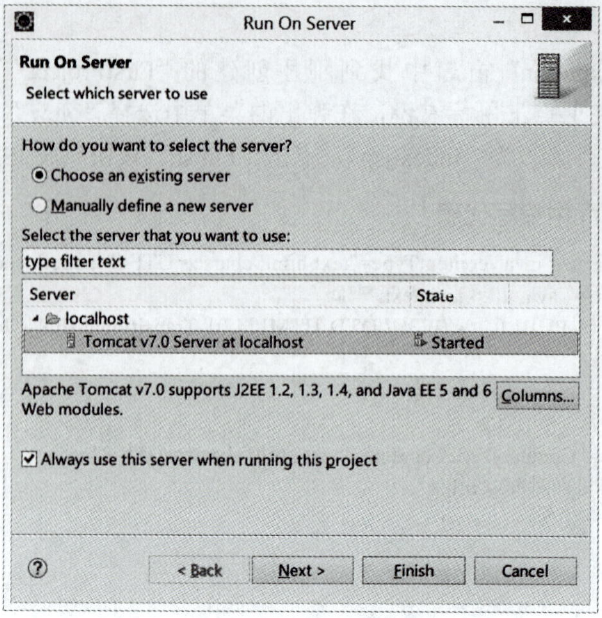

图 1.18 选择运行 Web 应用的服务器

拓展训练

1. 当 Eclipse 中的 Web 应用程序项目成功运行后，Tomcat 安装目录下的 server.xml 发生了什么变化（注意观察 Context 结点）？

2. Web 应用程序项目中的文件，在运行前被部署到了 Tomcat 下面的哪个目录中？项目中的内容和部署后的内容有什么对应关系？

同步训练

一、填空题

1. Tomcat 监听客户端请求的默认端口号是 _____。
2. Tomcat 的主要配置文件是 _____ 和 _____。
3. 为了在不重启 Tomcat 的情况下更新 Web 应用程序，需要将配置文件中 _____ 元素的 _____ 属性设置为 true。
4. Eclipse 的设置窗口通过点击菜单项 _____ 下的子菜单 _____ 打开。
5. Eclipse 的启动参数配置文件是 _____。

二、简答题

1. Tomcat 中最常用的配置文件是哪两个？存放在什么位置？

2．在 Tomcat 服务器的组织结构中，Server、Service、Connector、Engine、Host、Context 分别表示什么含义？其中 Engine、Host、Context 之间存在什么关系？

3．Tomcat 中默认的 Engine、Host 叫什么名字？该 Host 对应的目录名是什么？

4．什么是 Context？Host 中默认的 Context 存放在哪个目录中？

5．WEB-INF 是 Context 中的一个特殊目录，该目录下通常存放哪些内容？

6．如何设置一个 Context 的默认主页？

项目 2
实现用户注册功能

单元介绍

通过本项目，掌握网页表单标签的用法，以及通过表单提交数据后的接收、处理和响应。

本项目包括以下任务：
- 设计用户注册表单
- 接收表单提交的数据
- 验证提交的数据是否合法
- 处理表单提交的数据并响应输出

学习目标

【知识目标】
了解 HTTP 规范
掌握表单标签的特点、使用方法和注意事项
掌握接收表单提交数据的方法
掌握提交数据的处理和响应输出到客户端的方法

【能力目标】
能设计规范、正确的网页表单
能编写代码接收表单提交的数据
能编写代码将提交数据的处理和响应输出到客户端

任务 2.1　设计注册信息输入页面

【任务分析】

用户注册时提交信息的类型比较多，包括文本类型、数值类型、日期类型等，有的信息还有采用选择的方式，从指定选择中选择一项或多项。为了降低难度，注册页面中的表单输入元素都采用标准的 HTML 标签来实现。

【相关知识】

2.1.1　HTTP

HTTP 是 HyperText Transfer Protocol（超文本传输协议）的缩写，定义了客户端（如浏览器）和 Web 服务器之间的数据交换格式。HTTP 是一种无状态协议，客户端连接服务器后，发送请求数据，服务器端接收请求数据，处理后将结果返回给客户端，然后客户端和服务器端断开连接。HTTP 通信过程就是一个"请求（request）"—"响应（response）"的过程。

① HTTP 的请求报文格式

HTTP 请求报文组成部分包括：

- 初始行（initial line）
- 请求头（header lines）
- 空行（blank line）
- 消息正文（body）

一个用 HTTP 抓包工具抓取到的请求报文内容如下：

POST http://localhost:8080/FirstProject/login_check.jsp HTTP/1.1
Host: localhost:8080
Connection: keep-alive
Content-Length: 64
Cache-Control: max-age=0
Accept: text/html,application/xhtml+xml,application/xml;q=0.9,image/webp,*/*;q=0.8
Origin: http://localhost:8080
Upgrade-Insecure-Requests: 1
User-Agent: Mozilla/5.0 (Windows NT 6.3; WOW64) AppleWebKit/537.36 (KHTML, like Gecko) Chrome/50.0.2661.94 Safari/537.36
Content-Type: application/x-www-form-urlencoded
Referer: http://localhost:8080/FirstProject/
Accept-Encoding: gzip, deflate

Accept-Language: zh-CN,zh;q=0.8
Cookie: JSESSIONID=CEDA40765BA64F9C1D34FD089DB17811

uid=abc&pwd=123456&sex=M&birthday=1990-5-1&interest=1&interest=2

② HTTP 的响应报文格式

HTTP 响应报文组成部分包括：

- 初始行（initial line）
- 响应头（header lines）
- 空行（blank line）
- 消息正文（body）

一个用 HTTP 抓包工具抓取到的响应报文内容如下：

HTTP/1.1 200 OK
Server: Apache-Coyote/1.1
Content-Type: text/html;charset=UTF-8
Content-Length: 401
Date: Mon, 12 Aug 2016 09:18:40 GMT

<!DOCTYPE html PUBLIC "-//W3C//DTD HTML 4.01 Transitional//EN" "http://www.w3.org/TR/html4/loose.dtd">
<html>
<head>
<meta http-equiv="Content-Type" content="text/html; charset=UTF-8">
<title> 注册验证 </title>
</head>
<body>
提交的用户名是：abc
 提交的密码是：123456
 提交的性别是：M
 提交的生日是：1990-5-1
 提交的爱好是：1 2
</body>
</html>

下面对几个常用 header lines 进行说明。

Host：用于请求报文，表示请求的主机地址。

Content-Type：用于响应报文，表示返回资源的文件类型，以 MIME 形式给出，如 text/html、image/jpg、image/gif 等。

Content-Length：用于请求或响应报文，表示资源的文件长度。

2.1.2 HTML

HTML 是 HyperText Markup Language（超文本标记语言）的缩写，用 HTML 编写的文档称为超文本文档。HTML 本身为纯文本格式，但是可以通过一些特定的标签表示图片、声音、超链接等非文本内容。

当浏览器发起请求时，服务器通过 HTTP 响应给浏览器的数据正文通常用 HTML 格式表示，浏览器接收到之后，再解析并呈现为用户看到的五彩缤纷的网页。

HTML 中包含了很多的标签，比如：

①在网页中显示 flower.jpg 文件所对应的图片，高度和宽度都为 200

```
<img src="flower.jpg" width="200" height="200">
```
②在网页中显示一个超链接
```
<a href="http://www.baidu.com"> 百度 </a>
```
③在网页中显示一个表格
```
<table width="200px" border="1">
<tr><td>1 行 1 列 </td><td>1 行 2 列 </td></tr>
<tr><td>2 行 1 列 </td><td>2 行 2 列 </td></tr>
</table>
```

2.1.3 表单标签

当浏览器向服务器发送比较复杂的数据时，人工去构造 HTTP 请求报文是一项艰巨的任务，并且用户需要一个友好的界面来填写数据，而并不关心浏览器在后台是怎么把数据提交给服务器的。

HTML 的表单标签包括一系列的控件，以友好的图形界面供用户进行输入或选择，当点击"提交"按钮时，将数据封装为 HTTP 请求报文发送给服务器。表单标签包括的控件有：

① input（输入）标签

input 标签是网页中变化最多的标签，根据 type 属性不同，可以产生 10 种变化，表 2.1 中进行了归纳。

表 2.1 input 标签的 type 属性

type 属性取值	输入区域类型	示例
`<input type="TEXT" size="10" maxlength="8">`	单行文本输入区域，size 属性定义显示尺寸大小，maxlength 属性定义最大输入字符数	用户名：admin
`<input type="PASSWORD">`	输入密码的区域，当用户输入密码时，区域内将会显示"*"号	密码：*****
`<input type="SUBMIT" value=" 提交按钮 ">`	提交按钮，点击后将表单内容提交给服务器，value 属性指定按钮表面文字	提交按钮
`<input type="IMAGE">`	功能与提交按钮相同，但表现样式为图片，用 src 属性定义图片地址	
`<input type="RESET" value=" 重置按钮 ">`	重置按钮，点击后将表单内容全部清除，重新填写，value 属性指定按钮表面文字	重置按钮
`<input type="BUTTON" value=" 普通按钮 ">`	普通按钮，需要自己编写脚本程序处理点击事件，value 属性指定按钮表面文字	普通按钮
`<input type="CHECKBOX" checked>`	复选框，checked 属性用来设置该复选框缺省时是否被选中	植物生长需要：☑阳光 ☑空气 ☑水分
`<input type="HIDDEN">`	隐藏区域，在浏览器中不可见，用来存放某些要传递给服务器的数据	

续表

type 属性取值	输入区域类型	示例
<input type="RADIO">	单选按钮，checked 属性用来设置该单选框缺省时是否被选中	你认为巴西队能夺冠吗？ ⊙ 能　○ 不能　○ 不确定
<input type="FILE">	用来选择一个需要上传的文件	[_____] 浏览...

使用 input 标签时注意：
- input 标签一般要设置 name 属性，这样提交到服务器后才能获取到值。
- 对于文本、密码输入框，可以通过 value 属性设置初始内容。
- 对于按钮，可以通过 value 属性设置表面的提示文字。
- 单选按钮和复选框的 value 属性在浏览器中没有效果，但是可以提交到服务器上。
- 对于同在一个分组的单选按钮，name 属性必须设置为相同的值。

② select（选择）标签

select 标签一般和 option 标签配合使用，实现网页中的下拉框和列表框，如表 2.2 所示。

表 2.2　select 标签

select 标签代码	浏览器显示结果
<form name="form1" action="vote.asp" method="post"> 请选择最喜欢的男歌星 :
 <select name="s1" size="1"> <option value="ldh" selected> 刘德华 </option> <option value="zxy"> 张学友 </option> <option value="gfc"> 郭富城 </option> <option value="lm"> 黎明 </option> </select> <form>	请选择最喜欢的男歌星： 刘德华 ▼ 刘德华 张学友 郭富城 黎明
<form name="form1" action="vote.asp" method="post"> 请选择最喜欢的女歌星 :
 <select name="s2" multiple size="4"> <option value="wf" selected> 王菲 </option> <option value="tz"> 田震 </option> <option value="ny"> 那英 </option> </select> <form>	请选择最喜欢的女歌星： 王菲 田震 那英

使用 select 标签时注意：
- select 标签的 size 属性决定它的外观，默认情况下 size 为 1，显示为下拉框，如果把 size 设置为大于 1，就显示为列表框。
- select 标签的内容如果要提交，就必须包含在 <form></form> 中，并且要设置 name 属性。
- 在 select 标签中，每个 <option></option> 标签对表示一个选项，默认选中项要加 selected 属性。

- 在列表框中用 multiple 属性实现多选。

③ textarea（文本域）标签

textarea 标签的功能和 <input type="text"> 标签相似，都是实现文本输入，不同的是 textarea 可以输入较多内容，并且可以换行，如表 2.3 所示。

表 2.3 textarea 标签

textarea 标签代码	浏览器显示结果
<form name="form1" action="info.asp" method="post"> 请输入个性化签名： <textarea name="qm" clos="20" rows="5"> 此君很懒，什么都没有留下。 </textarea> </form>	请输入个性化签名： 此君很懒，什么都没有留下。

使用 textarea 标签时注意：

- textarea 标签必须成对使用。
- 可以用 cols、rows 属性设置文本域的列数（宽度）和行数（高度）。

④ form（表单）标签

<form></form> 标签对定义表单的开始和结束，上述的表单控件只有包含在标签对里面，其数据才能提交。常用的属性有 name、action、method、target 和 enctype，它们的含义如下：

- name，表单的名字。
- action，表单提交后处理程序的 url 地址。
- method，提交表单的方式，采用 get 方式提交，提交的内容会显示在浏览器地址栏中，提交的数据量很少；采用 post 方式提交，提交的内容不会显示在地址栏中，提交的数据量大。
- target，显示返回结果页面的目标窗口（和 a 标签的 target 属性类似）。
- enctype，上传数据的编码格式，上传文件一般用 "multipart/form-data"。

例如：

<form name="form1" action="reg_check.jsp" method="post" target="_blank">
</form>

这段代码表示表单的名字叫 form1，提交之后由 reg_check.jsp 处理，采用 POST 方式提交，提交后的处理结果页面显示在新窗口中。

2.1.4 CSS 定位与 DIV 布局

DIV+CSS 是 Web 设计标准，它是一种网页的布局方法，与传统的通过表格（table）布局定位的方式不同，它可以实现网页页面内容与表现相分离。

① CSS 盒子模型

CSS 控制页面是通过盒子模型实现的，所有页面标签都可以看成一个盒子，分布在页面的固定位置。盒子的大小需要通过属性调整，盒子和盒子之间的影响也通过浮动和定位等技术实现。

CSS 盒子模型如图 2.1 所示。

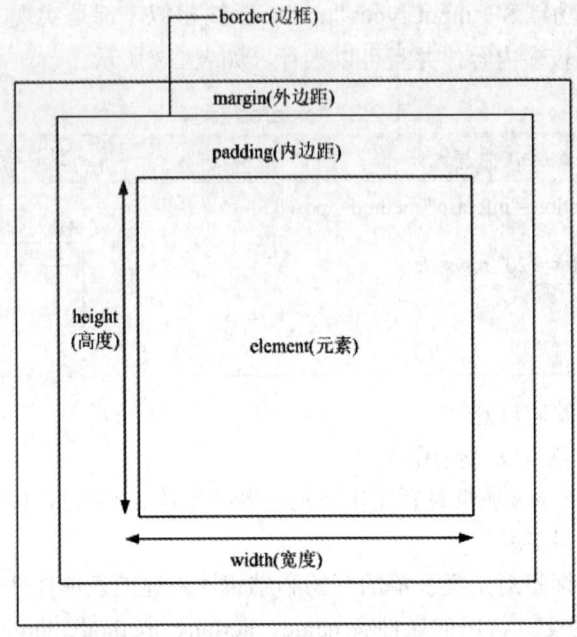

图 2.1　CSS 盒子模型图

从图 2.1 中可以看出，CSS 盒子模型包括 6 个部分，分别为元素内容（element）、宽度（width）、高度（height）、内边距（padding）、边框（border）和外边距（margin）。

- 元素内容（element）是指页面的实际内容。
- 宽度（width）和高度（height）主要用于控制 CSS 盒子模型中元素的大小，盒子的实际宽度 = 元素宽度 + 内边距（左侧和右侧内边距之和）+ 边框（左侧和右侧边框宽度之和）+ 外边距（左侧和右侧外边距之和），盒子的实际高度 = 元素高度 + 内边距（上侧和下侧内边距之和）+ 边框（上侧和下侧边框长度之和）+ 外边距（上侧和下侧外边距之和）。
- 内边距（padding）定义元素边框与元素内容之间的空白区域。该属性可以设置 1~4 个属性值。

当指定 1 个属性值时，表示上、下、左、右的内边距均为该值。
当指定 2 个属性值时，分别对应上、下和左、右的内边距。
当指定 3 个属性值时，分别表示上、左右、下的内边距。
当指定 4 个属性值时，分别表示上、下、左、右的内边距。
也可以通过 padding-top、padding-right、padding-bottom 和 padding-left 分别设置上、右、下、左内边距。

- 边框（border）是围绕元素内容和内边距的一条或多条线。该属性允许用户规定元素边框的样式、宽度和颜色。边框样式是通过 border-style 属性实现的，该属性可以设置为 none、hidden、dotted 等，也可以通过 top-right-bottom-left 的顺序为上、右、下、左边框指定样式，还可以通过 border-top-style、border-right-style、border-bottom-style 和 border-left-style 分别设置上、右、下、左边框样式。
- 外边距（margin）是指围绕在元素边框外的空白区域，设置外边距的最简单的方法就是使用 margin 属性，margin 属性接受任何长度单位，可以是像素、英寸、毫米或 em，margin 可以设置为 auto，margin 的默认值是 0，所以如果没有为 margin 声明一个值，就不会出现外边距。CSS 定义了一些规则，允许为外边距指定少于 4 个值，如果缺少左外边距的值，则使用右外边距的值，如果缺少下外边距的值，则使用上外边距的值，如果缺少右外边距的值，则使用上外边距的值。也可以通过 margin-top、margin-right、margin-bottom 和 margin-left 分别设置上、右、下、左外边距。

下面通过一个示例演示 CSS 盒子模型各个属性的用法，该示例实现一个背景为灰色的简单 CSS 盒子模型，具体代码如下。

```html
<!DOCTYPE html>
<html>
<head>
<title>CSS 盒子 </title>
<style type="text/css">
.box {
    background-color: #808080;  /* 指定背景颜色 */
    margin: 20px;           /* 指定外边距 */
    padding: 20px;          /* 指定内边距 */
    height: 60px;           /* 指定高度 */
    width: 100px;           /* 指定宽度 */
}
</style>
</head>
<body>
  <div class="box">
     我是一个 CSS 盒子模型
  </div>
</body>
</html>
```

程序的运行结果如图 2.2 所示。

图 2.2　CSS 盒子模型

② CSS 盒子的定位与浮动
- float 浮动

在标准流中，一个块级元素在页面中独占一行，各个块级元素自上而下排列。但使用了浮动后，块级元素的排列方式就会有所变化。

float 浮动属性可以指定 left、right、none 和 inherit。

left：元素向左浮动。

right：元素向右浮动。

none：默认值。元素不浮动，并会显示其在文本中出现的位置。

inherit：规定应该从父元素继承 float 属性的值。

下面通过一个示例来演示 float 浮动的效果。首先定义图 2.2 中所示的两个 CSS 盒子模型，具体代码如下：

```
<!DOCTYPE html>
<html>
<head>
<title>CSS 盒子 </title>
<style type="text/css">
.box {
    background-color: #808080;  /* 指定背景颜色 */
    margin: 20px;        /* 指定外边距 */
    padding: 20px;       /* 指定内边距 */
    height: 60px;        /* 指定高度 */
    width: 100px;        /* 指定宽度 */
}
</style>
</head>
<body>
  <div class="box">
     我是一个 CSS 盒子模型
  </div>
  <div class="box">
     我是一个 CSS 盒子模型
  </div>
</body>
</html>
```

程序的运行结果如图 2.3 所示。

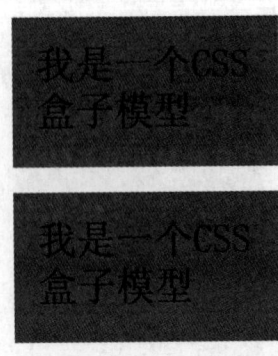

图 2.3　未使用浮动的 CSS 盒子模型

下面在样式中加入向左浮动效果，具体代码如下。

```html
<!DOCTYPE html>
<html>
<head>
<title>CSS 盒子 </title>
<style type="text/css">
.box {
  background-color: #808080;  /* 指定背景颜色 */
  margin: 20px;               /* 指定外边距 */
  padding: 20px;              /* 指定内边距 */
  height: 60px;               /* 指定高度 */
  width: 100px;               /* 指定宽度 */
  float: left;                /* 向左浮动 */
}
</style>
</head>
<body>
  <div class="box">
    我是一个 CSS 盒子模型
  </div>
  <div class="box">
    我是一个 CSS 盒子模型
  </div>
</body>
</html>
```

程序的运行结果如图 2.4 所示。

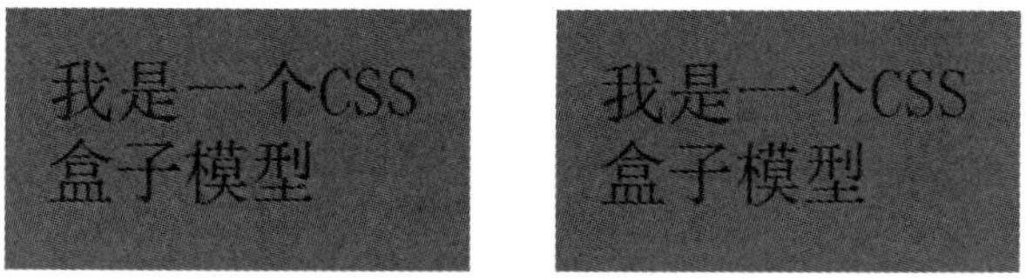

图 2.4　使用浮动的 CSS 盒子模型

通过上述两个示例，可以看出两个盒子由垂直排列变成了水平排列。

如果要清除浮动效果，需要使用 clear 属性，它的取值可以为 left、right、both、none 和 inherit。

left：在左侧不允许浮动元素。

right：在右侧不允许浮动元素。

both：在左、右两侧均不允许浮动元素。

none：默认值，允许浮动元素出现在两侧。

inherit：规定应该从父元素继承 clear 属性的值。

● position 定位

position 属性规定元素的定位类型。position 包含以下 5 个属性。

absolute：生成绝对定位的元素，相对于 static 定位以外的第一个父元素进行定位。

fixed：生成绝对定位的元素，相对于浏览器窗口进行定位。
relative：生成相对定位的元素，相对于其正常位置进行定位。
static：默认值。没有定位，元素出现在正常的流中。
inherit：规定应该从父元素继承 position 属性的值。

下面通过示例讲解这几种属性的区别，该示例实现一个含有嵌套关系的盒子。具体代码如下。

```html
<!DOCTYPE html>
<html>
<head>
<title>CSS 嵌套盒子 </title>
<style type="text/css">
.father {
  background-color: #D3D3D3;  /* 指定背景颜色 */
  height: 80px;           /* 指定高度 */
  width: 100px;           /* 指定宽度 */
}
.son {
  background-color: #696969;  /* 指定背景颜色 */
  height: 40px;           /* 指定高度 */
  width: 50px;            /* 指定宽度 */
}
</style>
</head>
<body>
  <div class="father">
    <div class="son">
      子盒子
    </div>
    父盒子
  </div>
</body>
</html>
```

程序的运行结果如图 2.5 所示。

图 2.5　嵌套盒子模型

如果想要让子盒子向右下各偏移 20px，使用相对定位的代码如下。

```css
.son {
```

```
  background-color: #696969;   /* 指定背景颜色 */
  height: 40px;                /* 指定高度 */
  width: 50px;                 /* 指定宽度 */
  position:relative;           /* 相对定位 */
  left:20px;                   /* 向右偏移 20 */
  top:20px;                    /* 向下偏移 20 */
}
```
程序的运行结果如图 2.6 所示。

图 2.6　使用相对定位实现嵌套盒子模型

相对定位是相对于原来的位置，通过偏移指定的距离到达新位置，而父标签中的其他标签不受影响。

如果使用绝对定位，代码如下。

```
.son {
  background-color: #696969;   /* 指定背景颜色 */
  height: 40px;                /* 指定高度 */
  width: 50px;                 /* 指定宽度 */
  position:absolute;           /* 绝对定位 */
  left:20px;                   /* 向右偏移 20 */
  top:20px;                    /* 向下偏移 20 */
}
```
程序的运行结果如图 2.7 所示。

图 2.7　使用绝对定位实现嵌套盒子模型

绝对定位以父标签为基准进行偏移，如果没有父标签，会以浏览器窗口为基准进行偏移，其他标签按照原有布局进行排列。

● display 属性

display 属性规定元素应该生成的显示框的类型，display 的值可以是 none、block

或 inline，none 表示此元素不会被显示；block 表示此元素将显示为块级元素，此元素前后会带有换行符；inline 为默认值，此元素会被显示为内联元素，元素前后没有换行符。

将块级标签 <div> 变成行内标签的代码如下：

 将块级标签变成行内标签

将行内标签 <div> 变成块级标签的代码如下：

 将行内标签变成块级标签

将块级标签和行内标签隐藏的代码如下：

 将块级标签隐藏
 将行内标签隐藏

【任务实施】

1. 新建项目

选择菜单"File"—"New"—"Dynamic Web Project"（如果没有列出，在"Other…"中查找），在弹出的窗口中填写"Project name"为"UserReg"，连续点击"Next"两次，在"Web Module"选项卡中勾选"Generate web.xml deployment descriptor"复选框，让项目中自动添加 web.xml 文件，最后点击"Finish"按钮创建新项目。

在 web.xml 中修改 Web 应用程序的主页设置，将主页改为 reg.jsp：

```
<welcome-file-list>
    <welcome-file>reg.jsp</welcome-file>
</welcome-file-list>
```

2. 创建 JSP 页面

在"Package Explorer"中展开"UserReg"项目，在 WebContent 文件夹上点击鼠标右键，选择"New"—"Other…"，在弹出的窗口中搜索"jsp"，如图 2.8 所示。

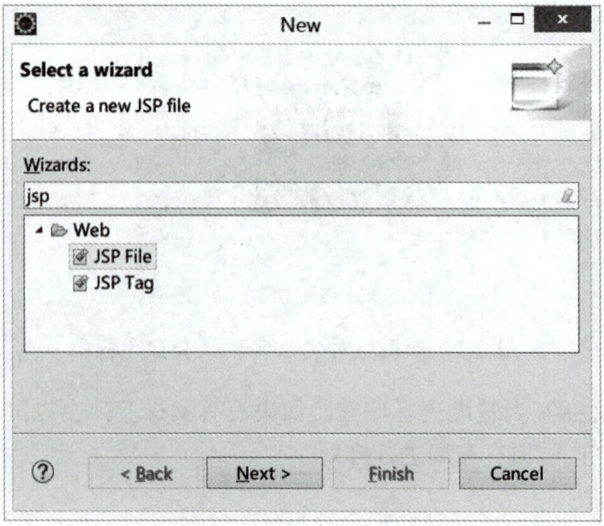

图 2.8　新建 JSP 文件

选择"JSP File",点击"Next>"按钮,将文件名改为"reg.jsp",再点击"Next>",在 JSP 文件模板中选择"New JSP File(html)",如图 2.9 所示。

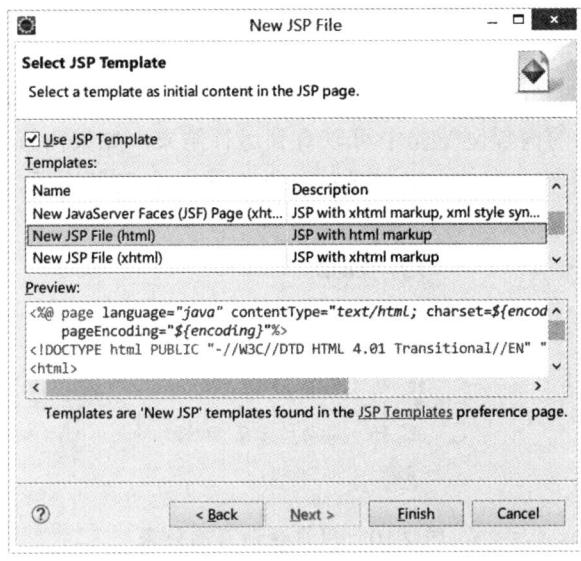

图 2.9 选择 JSP 文件模板

最后点击"Finish"完成 JSP 文件的创建。

3. 编辑 JSP 页面

打开 reg.jsp,输入网页内容:

<%@ page language="java" contentType="text/html; charset=UTF-8" pageEncoding="UTF-8"%>
<!DOCTYPE html PUBLIC "-//W3C//DTD HTML 4.01 Transitional//EN" "http://www.w3.org/TR/html4/loose.dtd">
<html>
<head>
<meta http-equiv="Content-Type" content="text/html; charset=UTF-8">
<title> 用户注册 </title>
</head>
<body>
<form action="reg_check.jsp" methid="post">
<h2> 用户注册 </h2>
用户名 <input type="text" name="uid">
密　码 <input type="password" name="pwd">
性　别 <select name="sex">
　　<option value="M"> 男 </option>
　　<option value="F"> 女 </option>
　</select>
生　日 <input type="text" name="birthday">
爱　好
<input type="checkbox" name="interest" value="1"> 运动
<input type="checkbox" name="interest" value="2"> 音乐
<input type="checkbox" name="interest" value="3"> 旅游


```
<input type="submit" value=" 注册 ">
</form>
</body>
</html>
```

4．运行项目

点击工具栏上的"Run"按钮（快捷键 Ctrl+F11），在弹出的窗口中选择"Run On Server"，在 Eclipse 的内置浏览器中可以看到运行结果，如果内置浏览器中没有出现，可以在外部浏览器中输入"http://localhost:8080/UserReg/"打开页面，效果如图 2.10 所示。

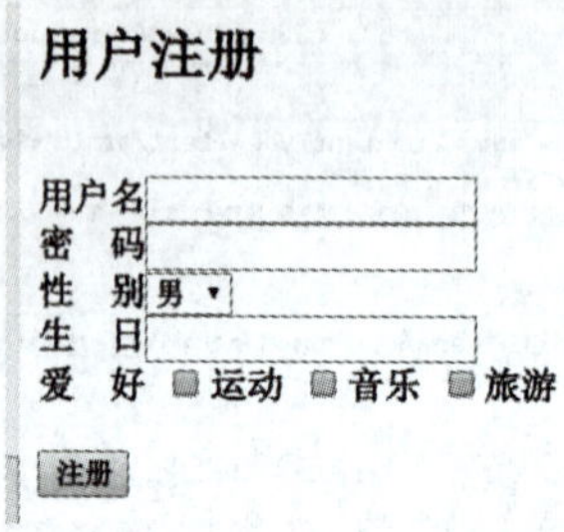

图 2.10　用户注册页面效果

【实践训练】

新建名为 UserReg 的 Web 应用项目，将默认主页设置为 reg.jsp，在 reg.jsp 中编写用户注册表单，并运行项目测试效果。

任务 2.2　处理表单提交的数据

【任务分析】

任务 2.1 完成了表单设计。当点击按钮提交表单时，浏览器将数据打包成 HTTP 请求报文发送到服务器，服务器端的处理程序通过 JSP 的内置对象 request 获取数据并进行处理，处理完毕后，再通过内置对象 Response 将结果打包成 HTTP 响应报文发送给浏览器。

【相关知识】

2.2.1　JSP 的运行原理

当 HTTP 请求被发送到指定的 JSP 页面时，Web 容器（如 Tomcat）会通过 3 个步骤来执行 JSP 页面：

① 将 JSP 页面内容翻译成 Java 源代码，生成一个后缀为 .java 的文件；

② 编译 Java 代码，生成后缀为 .class 的可执行文件；

③ 执行 .class 文件并向浏览器输出结果。

JSP 页面第一次被访问时，要从步骤①开始执行，所以速度比较慢；后续访问时，如果 JSP 页面没有发生改变，就直接执行步骤③，速度很快，如果访问的 JSP 页面内容被更改，Web 容器会重新对 JSP 页面进行翻译、编译和执行。

在 Tomcat 的根目录下，打开 "\work\Catalina\localhost\ 项目名称 \org\apache\jsp" 目录，可以看到由 JSP 文件翻译成的 Java 文件，如 "index.jsp" 被翻译成 "index_jsp.java"，通过对比这两个文件，可以分析 JSP 转换成 Java 的规律。

2.2.2 JSP 页面的组成部分

在实际动手编写 JSP 页面之前，先要了解页面的组成部分，以及各部分的作用。JSP 页面由 HTML、指令、Java 代码块、表达式、声明、注释等元素构成，如表 2.4 所示。

表 2.4　JSP 页面的组成部分

页面元素		说明
HTML		包括 HTML 标签、CSS 样式、JavaScript 脚本等静态内容，这部分内容是固定不变的。在将 JSP 翻译成 Java 源代码阶段，这部分内容的每一行将用 out.write 原样输出
JSP 指令		用来设置 JSP 页面的相关属性，只是告诉引擎如何处理 JSP 页面，并不直接产生任何可见的输出
JSP 动作标签		用来引入现有的文件或者控制 JSP 引擎的行为
嵌入的 Java 代码	Java 代码块	在将 JSP 翻译成 Java 源代码阶段，Java 代码块中的代码会放入 _jspService 方法（等价于 doGet 和 doPost）的方法体中，在接收、处理浏览器请求时执行
	Java 表达式	在将 JSP 翻译成 Java 源代码阶段，该表达式会通过 out.write 输出，即变成 "out.write（表达式）" 的形式
	Java 类成员声明	在将 JSP 翻译成 Java 源代码阶段，声明的变量称为类的数据成员，声明的函数称为类的方法成员，与 _jspService 方法平级
注释	HTML 注释	被注释的内容要发送给浏览器，但被浏览器当做 HTML 注释而不显示出来
	JSP 注释	服务器端在转换成 Java 代码时直接忽略注释内容
	Java 注释	只用于嵌入的 Java 代码中，服务器端在转换成 Java 代码时保留注释内容，但不会编译和执行

① JSP 指令

JSP 指令主要包括 page 指令和 include 指令。

● page 指令

page 指令用于定义页面的各种属性，一个 JSP 页面中可以包含多个 page 指令，但是指令属性除了 import 外，其他属性每个只能定义一次，否则翻译成 Java 源文件时会出现语法错误。page 指令的语法格式如下：

<%@page 属性 1=" 属性值 " 属性 2=" 属性值 " ... %>

page 指令的常用属性如表 2.5 所示。

表 2.5　page 指令的常用属性

属性	说明	举例
language	指定 JSP 中脚本的编程语言，目前只能为 Java	language="Java"
import	引入脚本中使用的类	import="java.util.*,java.sql.*"
contentType	指定 JSP 页面的内容格式和字符编码方式	contentType="text/html;charset=UTF-8"

- include 指令

include 指令在标签位置处静态包含一个文件，静态包含是将被包含文件的内容取出后代替标签内容，与当前 JSP 文件合并在一起，再翻译成 Java 源文件。语法格式如下：

<%@include file=" 被包含文件名 " %>

② JSP 动作标签

JSP 动作标签用来执行一些具体功能，起到简化 Java 代码的作用，常用的有 include、forward、useBean 标签。

- include 动作标签

include 动作标签在当前 JSP 页面中动态包含一个文件，即被包含的文件独立翻译为 Java 源代码并编译成独立的字节码文件（.class），当前 JSP 文件执行到 include 标签处时，才去加载被包含文件的字节码文件。语法格式如下：

<jsp:include page=" 被包含的文件名 "/>

- forward 动作标签

forward 动作标签作用是当执行到这里时跳转到另一个 JSP 页面执行，语法格式是：

<jsp:forward page=" 要跳转到的页面 ">

- useBean 动作标签

useBean 动作标签在 JSP 页面中根据指定的类名创建一个 Java 对象，并且可以设置该对象的生命周期。语法格式如下：

<jsp:useBean id=" 对象引用名 " scope=" 对象作用域 " class=" 包名 . 类名 "/>

其中 scope 属性的可选择值如表 2.6 所示。

表 2.6　scope 属性的可选值

值	说明
page	对象只在当前页中有效（默认值）
request	对象在请求的生命周期内有效（一次请求可以跨越多个 JSP 页面），相当于用 request.setAttribute 将对象存入 Request 对象中

续表

值	说明
session	对象在用户会话的生命周期内有效，相当于用 session.setAttribute 将对象存入 Session 对象中
application	对象在整个 Web 应用程序运行期间都有效，相当于用 application.setAttribute 将对象存入 Session 对象中

③嵌入的 Java 代码

● Java 代码块

在 JSP 页面中，Java 代码块被包含在"<%"和"%>"之间，在代码中可以使用 JSP 的内置对象，也可以调用外部引入的类（通过 page 指令引入），例如：

```
<%
Date date = new Date();
    SimpleDateFormat df = new SimpleDateFormat("yyyy-MM-dd HH:mm:ss");
String str = df.format(date);
out.print(" 现在时间是：" + str);
%>
```

● Java 表达式

在 JSP 页面中，Java 表达式被包含在"<%="和"%>"之间，用于输出表达式的值，作用和 out.print 相同，但是可以融入 HTML 标签中。上面例子中输出时间语句可以用 Java 表达式表示为：

```
<%
Date date = new Date();
    SimpleDateFormat df = new SimpleDateFormat("yyyy-MM-dd HH:mm:ss");
String str = df.format(date);
%>
<h1><%=" 现在时间是：" + str %><h1>
```

● Java 类成员声明

声明为类成员的代码包含在"<%!"和"%>"中，可以声明类的数据成员和方法成员，例如：

```
<%!
private String formatDate(Date date)
{
  SimpleDateFormat df = new SimpleDateFormat("yyyy-MM-dd HH:mm:ss");
  return df.format(date);
}
%>
<h1><%=" 现在时间是：" + formatDate(new Date())%><h1>
```

代码中定义了一个名为 formatDate 的方法，并在 Java 表达式中调用。

④注释

● HTML 注释

HTML 注释包含在"<!--"和"-->"中，并且在"<%"和"%>"外，浏览器能接收到 HTML 源代码，但是不显示，例如：

```
<!--HTML 注释内容 -->
```
- JSP 注释

JSP 注释包含在"<%--"和"--%>"之间，注释的内容在 JSP 转换为 Java 阶段会被忽略，不会发送到客户端，例如：

```
<%--JSP 注释内容 --%>
```
- Java 注释

Java 注释只能包含在"<%"和"%>"之中，对里面的 Java 代码进行单行注释或多行注释，被注释的 Java 代码不会执行，例如：

```
<%
    // 单行注释
    /*
       多行注释
    */
%>
```

2.2.3　JSP 内置对象

JSP 中共预定义了 9 个对象，分别为：request、response、session、application、out、pageContext、config、page、exception，这些对象不需要程序员编写代码创建即可拿来使用，故称为内置对象。

其中 request、session、application 中包含了一个 HashMap，以"键值对"集合的方式存放数据，可以用 setAttribute 方法存入一条数据，用 getAttribute 方法取出一条数据，用 removeAttribute 删除一条数据。例如，void removeAttribute(String key) 删除关键字 key 对应的属性值。

① request 对象

request 对象所属的类实现了 javax.servlet.http.HttpServletRequest 接口，主要用于接收通过 HTTP 传输到服务器的数据，该对象的作用域为一次请求。

request 对象的常用方法有：

String getParameter(String name) 获取表单提交的信息。

String getProtocol() 获取客户端使用的协议。

String getServletPath() 获取客户端请求的页面路径。

String getMethod() 获取客户端请求的信息提交方式，如 GET、POST。

String getHeader() 获取 HTTP 请求头。

String getRemoteAddr() 获取客户端的 IP 地址。

String getRemoteHost() 获取客户端主机名称。

String getServerName() 获取服务器名称。

int getServerPort() 获取服务器端口号。

Enumeration<String> getParameterNames() 获取客户端提交的所有参数的名称。

void setAttribute(String key,Object val) 设置关键字 key 对应的属性值。

Object getAttribute(String key) 获取关键字 key 对应的属性值。

void removeAttribute(String key) 删除关键字 key 对应的属性值。

Enumeration<String> getAttributeNames() 返回 request 对象所有属性的关键字集合。

② response 对象

response 所属的类实现了 javax.servlet.http.HttpServletResponse 接口，对客户端请求进行响应，将 JSP 容器处理过的数据传回客户端。response 对象只在 JSP 页面内有效。常用的方法有：

String getCharacterEncoding() 获取响应字符编码格式。

void setCharacterEncoding(String charset) 设置响应字符编码格式。

void setContentType(String type) 设置响应 MIME 类型。

String sendRedirect(String location) 重定向（跳转）到另外一个页面。

PrintWriter getWriter() 获取打印输出对象。

③ session 对象

session 对象所属的类实现了 javax.servlet.http.HttpSession 接口，它表示客户端和服务器的一次会话。session 对象从一个客户打开浏览器并连接到服务器开始，到客户关闭浏览器离开这个服务器结束。浏览器在这个服务器的几个页面之间切换时，服务器端程序通过 session 对象来认定这是同一个客户，每个用户都拥有自己的 session。

session 的具体的工作过程是：当一个客户首次访问服务器上的 JSP 页面时，服务器端产生一个 session 对象存起来，同时分配一个标识该对象的 ID（字符串），并将这个 ID 发送到客户端，存放在浏览器的 cookie 中，直到客户关闭浏览器后，服务器端存储的该客户的 session 对象才被释放，和客户的会话结束。当客户重新打开浏览器再连接到该服务器时，服务器为该客户再创建一个新的 session 对象，开始一轮新的会话。

session 对象的常用方法有：

String getId() 获取 session 对象 ID。

void setAttribute(String key,Object obj) 设置关键字 key 对应的属性值。

Object getAttribute(String key) 获取关键字 key 对应的属性值。

void removeAttribute(String key) 删除关键字 key 对应的属性值。

public boolean isNew() 判断是否为一次新的会话。

④ application 对象

application 对象所属的类实现了 javax.servlet.ServletContext 接口，用来存储和共享一些全局可用的数据，application 对象中保存的信息在整个 Web 应用运行期间都有效。

与 session 对象不同，application 对象为所有客户公用，即所有客户共享这个内置的 application 对象。

String getContextPath() 获取 Web 应用程序的目录。

void SetAttribute(String key, Object value) 设置关键字 key 对应的属性值。

void RemoveAttribute(String key) 删除关键字 key 对应的属性值。

Object getAttribute(String key) 获取关键字 key 对应的属性值。

⑤ out 对象

out 对象所属的类实现了 javax.servlet.jsp.jspWriter 接口，用来向客户端输出各种类型的数据。常用方法有：

void print(各种类型) 输出各种类型数据，包括 boolean、char、int、double、float、long、string、object 等。

void newLine() 输出一个换行符。

⑥ pageContext 对象

pageContext 对象属于 javax.servlet.jsp.PageContext 类，通过它可以获取 JSP 页面的 request、response、session、application、out 等对象。pageContext 对象的创建和初始化都由容器来完成，在 JSP 页面中可以直接使用 pageContext 对象。

⑦ config 对象

config 对象所属的类实现了 javax.servlet.ServletConfig 接口，作用是取得服务器的配置信息。通过 pageConext 对象的 getServletConfig() 方法可以获取一个 config 对象。

当一个 Servlet 初始化时，容器把配置信息通过 config 对象传递过来。开发者可以在 web.xml 文件中为 Web 应用程序中的 Servlet 程序和 JSP 页面提供初始化参数。

⑧ page 对象

page 对象所属的类实现了 javax.servlet.jsp.HttpJspPage 接口，代表当前的 JSP 页面，也代表当前 JSP 编译后的 Servlet 类的对象，好比 Java 类中的 this 指针。

⑨ exception 对象

exception 对象属于 java.lang.Throwable 类，对象中存放异常信息。如果在 JSP 页面中出现没有用 try-catch 语句捕获的异常情况，就会生成 exception 对象，并把 exception 对象传输到在 page 指令中设定的错误页面，然后在错误页面中处理该 exception 对象。

例如，在产生异常的页面中定义错误处理页面为 ShowError.jsp：

<%@ page errorPage="ShowError.jsp" %>

在 ShowError.jsp 中，定义该页面为错误处理页，并输出错误信息：

<%@ page isErrorPage="true" %>
<html>
<head>
<title>Error</title>
</head>
<body>
<h2> 页面出现错误 </h2>
<h3> 错误信息：
<%= exception.getMessage(); %>
</h3>

> **注意**
>
> Exception 对象只有在 page 指令中包含属性 isErrorPage="true" 的 JSP 页面中才可以使用。

⑩使用 cookie 长期存储信息

cookie 本意是"饼干",这里表示保存在浏览器缓存中的一些文本信息,它是 javax.servlet.http.Cookie 类的对象。服务器端通过 response 对象将 cookie 中包含的信息存储到浏览器缓存中,以后可以通过 request 对象取回。

在 JSP 中,若要保存 cookie,应先创建一个 Cookie 类的对象,再用 response 对象的 addCookie() 方法添加到集合中发送给浏览器。创建 cookie 时,Cookie 类的构造函数接受两个字符串类型参数:cookie 名称和 cookie 值。

若要获取 cookie,采用 request 对象的 getCookies() 方法,从浏览器发送来的数据中得到 Cookie 类的对象数组,再循环比较每个对象的名称,找到想要的 cookie。

cookie 具有时效性,用 Cookie 类的 setMaxAge() 方法可以设置 cookie 的有效时间,该方法定义如下:

Void setMaxAge(int expiry)

其中参数 expiry 为 cookie 的有效时间,以秒为单位。

2.2.4　表单提交数据的接收

浏览器将表单中的数据以 HTTP 请求报文形式提交给 Tomcat 服务器,由服务器端程序(Servlet)来处理。传统的 Servlet 是一个 HttpServlet 的派生类,在里面覆盖(Override)基类 HttpServlet 的 doGet,处理用 GET 方式提交的数据,或覆盖 doPost,处理用 POST 方式提交的数据。

下面的代码编写了一个名为 RegServlet 的类,继承自 HttpServlet,并覆盖 doGet 和 doPost 方法:

```
public class RegServlet extends HttpServlet {
  @Override  //Override 注解,表示覆盖基类 HttpServlet 的 doGet 方法
    protected void doGet(HttpServletRequest request, HttpServletResponse response) throws ServletException, IOException
  {
    // 在这里接收、处理 GET 方式提交的数据
  }
  @Override  //Override 注解,表示覆盖基类 HttpServlet 的 doPost 方法
    protected void doPost(HttpServletRequest request, HttpServletResponse response) throws ServletException, IOException
{
        // 在这里接收、处理 POST 方式提交的数据
  }
}
```

在 doGet 或 doPost 方法中,有 request 和 response 两个参数,其中 request 负责处理请求(接收),response 负责处理响应(回复)。requset 对象最常用的两个方法是:

● public String getParameter(String name)

getParameter 方法根据表单控件名称,获取该名称对应的单一值,对应的控件类型包括文本框、密码框、单选按钮、下列框、列表框(单选)、多行输入文本框,返回结果为字符串。

- public String[] getParameterValues(String name)

getParameterValues 方法根据表单控件名称，获取该名称对应的多个值，对应的控件类型包括复选框、列表框（多选），返回结果为字符串数组。

用 getParameter 或 getParameterValues 获取到的数据都是字符串形式，如果提交的是数值、日期等非字符串类型，在处理过程中需要先进行类型转换。

JSP 页面的本质是 Servlet，但是 JSP 对 Servlet 进行了简化，隐藏了继承 HttpServlet 并覆盖 doGet、doPost 等方法的过程，还可以和 HTML 结合在一起。在 JSP 页面中，可以直接使用内置对象 request 的 getParameter 或 getParameterValues 方法获取请求的数据：

```jsp
<%@ page language="java" contentType="text/html; charset=UTF-8" pageEncoding="UTF-8"%>
<!DOCTYPE html PUBLIC "-//W3C//DTD HTML 4.01 Transitional//EN" "http://www.w3.org/TR/html4/loose.dtd">
<html>
<head>
<meta http-equiv="Content-Type" content="text/html; charset=UTF-8">
<title> 获取请求数据 </title>
</head>
<body>
<%
String uid=request.getParameter("uid"); // 获取用户登录名
String[] interests=request.getParameterValues("interest"); // 获取用户的爱好
%>
</body>
</html>
```

2.2.5 验证输入信息

数据格式的验证比较复杂，工程实践中通常借助正则表达式工具来完成。正则表达式（Regular Expression）也叫规则表达式，由于用字符串描述规则，也称为规则字符串。正则表达式起源于 UNIX 系统，主要用于合法性检查、字符串搜索、替换等，由于功能非常强大，很多操作系统和开发工具都提供了支持。JDK 从 1.4 版本开始，正式提供了对正则表达式的支持，对应的包是 java.util.regex。下面通过一段 Java 代码了解正则表达式的用法：

```java
regexTest.java
import java.util.regex.*;
public class regexTest
{
  public static void main(String[] args)
  {
    String s="ab12345";
    Pattern p=Pattern.compile("^[A-Z].*[A-Z]$", Pattern.CASE_INSENSITIVE);
    Matcher m=p.matcher(s);
    System.out.println(m.find());
  }
}
```

程序中先建立了一个名为 p 的模板，该模板规定"以大写字母开头，以大写字母结尾"，Pattern.CASE_INSENSITIVE 表示忽略大小写，再调用模板的 matcher 方法检查字符串 s 是否与之匹配，由于"ab12345"不能与模板匹配，输出结果为 false。

字符串"^[A-Z].*[A-Z]$"是描述规则的关键，即"规则字符串"，规则字符串主要由匹配对象、限定符、定位符、管道符构成。

①匹配对象

● 基本字符

比如普通的字母、数字、汉字等。

● 特殊符号

特殊符号匹配满足条件的字符集，或敏感字符，如表 2.7 所示。

表 2.7 表示特殊匹配对象的符号

符号	含义	符号	含义
\s	匹配任何空白字符，包括空格、制表符、换页符等	.	匹配除换行符之外的所有字符
\S	匹配任何非空白字符，与 \s 相反	\uxxxx	匹配 Unicode 编码
\d	匹配一个数字字符，等价于 [0～9]	*	匹配 * 号
\D	匹配一个非数字字符	\?	匹配问号
\w	匹配字母、数字或下划线	\\	匹配斜线
\W	匹配所有与 \w 不匹配的字符		

● 范围

正则表达式允许使用者在匹配模式中指定某一个范围而不局限于具体的字符。例如：

[A-Z] 与 A～Z 范围内任何一个大写字母相匹配。

[a-z] 与 a～z 范围内任何一个小写字母相匹配。

[0-9] 与 0～9 范围内任何一个数字相匹配。

[a-zA-Z0-9] 匹配一个字母或数字。

②限定符

限定符规定匹配对象在被检测字符串中出现的次数，较为常用的限定符包括："+""*""?"。

● "+"规定其前导字符必须在目标对象中连续出现一次或多次。

● "*"规定其前导字符必须在目标对象中出现零次或连续多次。

● "?"规定其前导字符必须在目标对象中连续出现零次或一次。

例如：xy+ 规则 与 "xy" 与 "xyyy" 相匹配，[a-z]+ 表示小写字母连续出现一次或多次。

除了限定符之外，还可以精确指定模式在匹配对象中出现的频率。

例如：a{3} 规定字符 a 在被测字符串中出现 3 次，[a-z]{1,3} 规定小写字母在被测字符串中出现最少 1 次、最多 3 次。

③定位符

定位符用于规定匹配模式在目标对象中的出现位置，较为常用的定位符包括：

"^" "$" "\b" 以及 "\B"。
- "^" 规定匹配模式必须出现在目标字符串的开头。
- "$" 规定匹配模式必须出现在目标对象的结尾。
- "\b" 规定匹配模式必须出现在目标字符串的开头或结尾的两个边界之一。
- "\B" 规定匹配对象必须位于目标字符串的开头和结尾两个边界之内，即匹配对象既不能作为目标字符串的开头，也不能作为目标字符串的结尾。

"^" 和 "$" 以及 "\b" 和 "\B" 是互为逆运算的两组定位符。

例如：^hell 和 "hell" 和 "hello" 相匹配。

④管道符（逻辑运算符）
- 或运算（|）

如果我们希望在正则表达式中实现类似逻辑中的"或"运算，在多个不同的模式中任选一个进行匹配的话，可以使用或运算符"|"。

例如：规则 ten|10 与 "ten" 或 "10" 相匹配。
- 非运算（^）

非运算对规则进行否定，注意非运算符的用法。

当 "^" 出现在 "[]" 内时就被视做非运算符；而当 "^" 位于 "[]" 之外，或没有 "[]" 时，则应当被视做定位符。

例如：规则 [^A-Z] 表示与非大写字母匹配。

⑤运算符的优先级

正则表达式运算符的优先级如表 2.8 所示。

表 2.8 正则表达式运算符的优先级

操作符	描述
\	转义符
()、[]	圆括号和方括号
*、+、?、{n}、{n,m}	限定符
^、$、\B、\b	定位符
|	或运算

2.2.6 响应输出到浏览器

在 Servlet 的 doGet 或 doPost 方法中，通过 response 参数，可以向浏览器输出结果。如果采用 JSP 页面，则直接使用 out 对象的 print 方法将结果输出到浏览器中。访问下面的 out.jsp，浏览器中会输出"欢迎访问用 JSP 制作的网页！"。

out.jsp
<%@ page language="java" contentType="text/html; charset=UTF-8" pageEncoding="UTF-8"%>
<!DOCTYPE html PUBLIC "-//W3C//DTD HTML 4.01 Transitional//EN" "http://www.w3.org/TR/html4/loose.dtd">
<html>

```jsp
<head>
<meta http-equiv="Content-Type" content="text/html; charset=UTF-8">
<title>JSP 内置对象 out 演示 </title>
</head>
<body>
<%
out.print(" 欢迎访问用 JSP 制作的网页！ ");
%>
</body>
</html>
```

【任务实施】

1．接收表单提交的数据并原样输出

在 WebContent 文件夹中创建 reg_check.jsp，创建时模板中选择"New JSP File (html)"，该页面负责接收和处理提交的注册信息。内容如下：

reg_check.jsp

```jsp
<%@ page language="java" contentType="text/html; charset=UTF-8" pageEncoding="UTF-8"%>
<!DOCTYPE html PUBLIC "-//W3C//DTD HTML 4.01 Transitional//EN" "http://www.w3.org/TR/html4/loose.dtd">
<html>
<head>
<meta http-equiv="Content-Type" content="text/html; charset=UTF-8">
<title> 接收注册信息并原样输出 </title>
</head>
<body>
<%
request.setCharacterEncoding("UTF-8"); // 修改文本数据的解码方式（JSP 默认用 ISO-8859-1）
String uid=request.getParameter("uid");
String pwd=request.getParameter("pwd");
String sex=request.getParameter("sex");
String birthday=request.getParameter("birthday");
String[] interest=request.getParameterValues("interest");
out.print(" 提交的用户名是："+uid+"<br>");
out.print(" 提交的密码是："+pwd+"<br>");
out.print(" 提交的性别是："+sex+"<br>");
out.print(" 提交的生日是："+birthday+"<br>");
out.print(" 提交的爱好是：");
if(interest!=null) // 没有勾选任何一项时，数组为空
   for(String i : interest) out.print(i+" ");
%>
</body>
</html>
```

运行 Web 应用程序项目后，在注册界面（见图 2.11）填写相关信息，填写完毕后

点击"注册"按钮提交表单。

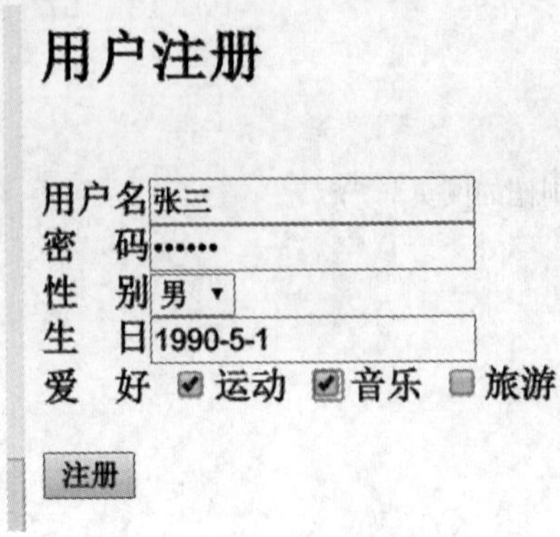

图 2.11　用户注册信息填写

提交给 reg_check.jsp 处理后，返回给浏览器的结果如图 2.12 所示。

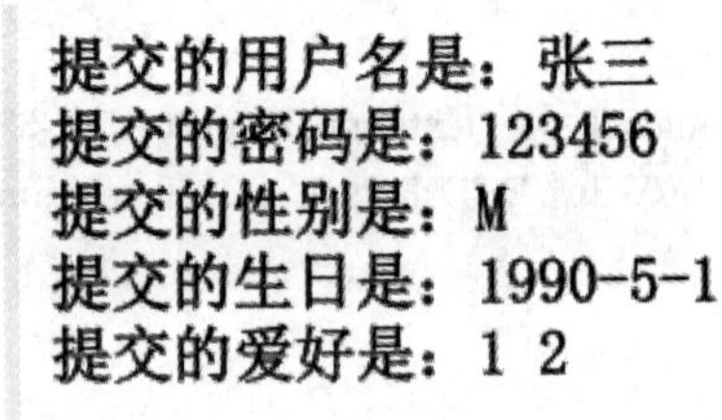

图 2.12　处理后的返回信息

2. 接收并验证表单提交的数据，输出验证结果

页面 reg_check2.jsp 对用户名、密码、生日这三项进行验证，其中用户名、密码用正则式验证，生日用日期格式化类验证：

reg_check2.jsp
```
<%@page import="java.text.SimpleDateFormat"%>
<%@page import="java.util.regex.*" %>
<%@page import="java.util.Date"%>
<%@page language="java" contentType="text/html; charset=UTF-8" pageEncoding ="UTF-8"%>
<!DOCTYPE html PUBLIC "-//W3C//DTD HTML 4.01 Transitional//EN" "http://www.w3.org/TR/html4/loose.dtd">
<html>
<head>
<meta http-equiv="Content-Type" content="text/html; charset=UTF-8">
<title> 接收注册信息并验证 </title>
<style type="text/css">
.error{color:red}
</style>
</head>
<body>
<%
```

```jsp
// 修改提交数据的解码方式（JSP 默认用 ISO-8859-1），解决提交中文数据乱码问题
request.setCharacterEncoding("UTF-8");
String uid=request.getParameter("uid");
String pwd=request.getParameter("pwd");
String sex=request.getParameter("sex");
String birthday=request.getParameter("birthday");
String[] interest=request.getParameterValues("interest");
out.print(" 提交的用户名是："+uid+"<br>");
Pattern p=Pattern.compile("^[A-Za-z0-9]{3,6}$");
if(!p.matcher(uid).find())
   out.print("<div class='error'> 用户名格式不合法（要求由 3-6 个字母和数字组成）</div>");
out.print(" 提交的密码是："+pwd+"<br>");
p=Pattern.compile("^[A-Za-z0-9]{6,10}$");
if(!p.matcher(pwd).find())
   out.print("<div class='error'> 密码格式不合法（要求由 6-10 个字母和数字组成）</div>");
out.print(" 提交的性别是："+sex+"<br>");
out.print(" 提交的生日是："+birthday+"<br>");
if(!isDate(birthday))
   out.print("<div class='error'> 生日格式不合法 </div>");
out.print(" 提交的爱好是：");
if(interest!=null)
   for(String i : interest) out.print(i+" ");
%>
<%!
// 函数定义在以 "<%!" 开头的代码块中，将被转换成 Servlet 的方法
boolean isDate(String s) // 检测一个字符串是否能转换成日期
{
   SimpleDateFormat df=new SimpleDateFormat("yyyy-MM-dd");
   df.setLenient(false); // 设置为严格验证
   try
   {
      df.parse(s);
      return true;
   }
   catch (Exception e)
   {
      return false;
   }
}
%>
</body>
</html>
```

将表单的 action 属性由 "reg_check.jsp" 改为 "reg_check2.jsp"，重新提交表单，观察数据验证结果，错误信息部分将在该项下方用红色表示。

【实践训练】

设计用户注册页面，并对提交的数据进行验证，要求如下：

（1）用户名长度为 3～8，密码长度为 6～10。
（2）用户名只能包含字母、数字和下划线，密码必须是字母和数字的组合。
（3）用户生日为合法的日期。

拓展训练

1. 查阅资料，了解 Fiddler 软件的用法，并用 Fiddler 抓取和分析 HTTP 报文。
2. 改进注册页面，注册信息增加手机号码和电子邮箱两项，并进行合法性验证。

同步训练

一、填空题

1. 表单的提交方式主要有 _____ 和 _____ 两种。
2. form 标签用 _____ 属性指定表单提交后的处理页面。
3. select 标签表现为下拉框和列表框的主要区别是 _____ 属性。
4. 能够多选的表单控件是 _____ 和 _____。
5. request 对象属于 _____ 类，response 对象属于 _____ 类。
6. 调用 request 对象的 _____ 方法设置接收数据编码方式。
7. 接收单值表单元素数据用 request 对象的 _____ 方法，接收多值表单元素用 request 对象的 _____ 方法。
8. 输出结果用 out 对象的 _____ 方法。

二、简答题

1. HTTP 请求报文和响应报文分别由哪几部分组成？
2. 什么是 HTML？HTML 代码本身是什么格式？
3. form 标签包含哪些主要属性？这些属性起什么作用？
4. 如何解决接收中文数据时的乱码问题？
5. 以下正则式的含义是什么？

① ^[A-Za-z]   ② ^[A-Za-z].*[A-Za-z]$ ③ ^[A-Za-z0-9]+$
④ ^[A-Za-z][A-Za-z0-9]+$ ⑤ ^[A-Za-z][A-Za-z0-9]{0,4}$

项目 3
改进用户注册功能

单元介绍

通过本项目，掌握 JDBC 的封装、数据库操作的分层设计、用 Servlet 接收和处理数据。

本项目包括以下任务：
- 使用 JDBC 连接及操作数据库
- 封装数据库的常用操作
- 编写 Servlet
- 在 JSP 中运用 EL 表达式

学习目标

【知识目标】

掌握 JDBC 常用的 API
了解常见数据库的 URL 格式
掌握数据库常用操作的封装及好处
掌握数据库操作分层设计理念及好处
掌握 Servlet 的编写步骤
掌握 EL 表达式

【能力目标】

能正确使用 JDBC 常用 API
能正确使用封装数据库常用操作
能编写数据库分层代码
能正确编写 Servlet
能正确使用 EL 表达式

任务 3.1　验证和保存用户信息

【任务分析】

将用户信息封装到实体类中，采用分层方式设计代码结构，并将用户信息保存到数据库中。在保存用户信息之前，先验证数据库中用户名是否存在，若存在，则进行提示；若不存在，则执行插入语句添加新用户。

【相关知识】

3.1.1　JDBC 的基本用法

将数据保存到数据库中能够实现长久的保存数据的目的。Java 程序通过 JDBC 和底层的数据库驱动程序，连接数据库并执行 SQL 语句来访问数据库中的数据。

① JDBC 概述

JDBC（Java Database Connectivity）是 Java 数据库连接的简称，由一组用 Java 语言编写的类和接口组成，用来实现对数据库的各种操作。为了实现各种数据库访问接口的统一，Java 提供了 JDBC 的应用程序接口规范——JDBC API，这些接口定义在 java.sql 包中，如 java.sql.Connection、java.sql.Statement、java.sql.PreparedStatement、java.sql.ResultSet 等。各个数据库提供商为该接口规范提供了具体的底层实现，即 JDBC 驱动。

② 常用的 JDBC API

常用的 JDBC API 包括 DriverManager、Connection、Statement、PreparedStatement、CallableStatement 和 ResultSet，JDBC API 可分为连接数据库的 API 和操作数据库的 API，连接数据库的 API 主要包括 DriverManager 和 Connection，操作数据库的 API 主要包括 Statement、PreparedStatement、CallableStatement 和 ResultSet。

● 连接数据库的 API

DriverManager 是管理 JDBC 驱动程序的基本服务，其中有一个重要的方法 getConnection()，调用该方法会返回 Connection 类型的引用。

DriverManager 中常用的方法如表 3.1 所示，其他方法可查阅 JDK 文档进行学习。

Connection 表示与特定数据库连接的会话，通过 DriverManager.getConnection() 方法获取。常用的方法如表 3.2 所示，其他方法可查阅 JDK 文档进行学习。

表 3.1　DriverManager 的常用方法

方法名	方法描述
public static Connection getConnection(String url, String user, String password) throws SQLException	获得 url 对应数据库的一个连接
public static int getLoginTimeout()	获取驱动程序登录到数据库时可以等待的最长时间，以秒为单位
public static void setLoginTimeout(int seconds)	设置驱动程序连接到数据库时等待的最长时间，以秒为单位

表 3.2　Connection 的常用方法

方法名	方法描述
Statement createStatement()throwsSQLException	创建一个 Statement 对象将 SQL 语句发送到数据库
CallableStatement prepareCall(String sql)throws SQLException	创建一个 CallableStatement 对象调用数据库存储过程
PreparedStatement prepareStatement(String sql) throws SQLException	创建 PreparedStatement 对象将参数化的 SQL 语句发送到数据库
void setAutoCommit(boolean autoCommit) throws SQLException	设置此连接的自动提交模式
void commit()throws SQLException	使所有上一次提交/回滚后进行的更改成为持久更改，此方法只能在已禁用自动提交模式时使用
void rollback()throws SQLException	取消在当前事务中进行的所有更改，此方法只能在已禁用自动提交模式时使用
void close()throws SQLException	立即释放此 Connection 对象的数据库和 JDBC 资源

● 操作数据库的 API

Statement 的主要作用是向数据库提交 SQL 语句并返回相应结果。执行的语句可以是插入、删除、查询和修改。常用的方法如表 3.3 所示，其他方法可查阅 JDK 文档进行学习。

表 3.3　Statement 的常用方法

方法名	方法描述
void close()throws SQLException	关闭 Statement 对象并立即释放此对象的资源
boolean execute(String sql) throws SQLException	执行一条 INSERT、UPDATE、DELETE 或不返回任何内容的 SQL 语句，返回值表示是否执行成功
ResultSet executeQuery(String sql) throws SQLException	执行给定的查询语句并将查询结果返回 ResultSet（结果集）对象中
int executeUpdate(String sql) throws SQLException	执行一条 INSERT、UPDATE、DELETE 语句，返回发生改变的记录条数

PreparedStatement 接口继承 Statement 接口，因此该类型的引用可以使用 Statement 中的所有方法。当一条语句需要修改参数而多次运行时，选择 PreparedStatement 能够提高执行效率。

PreparedStatement 允许 SQL 语句设置输入参数，使用"?"作为占位符。在获得 PreparedStatement 对象后，执行 SQL 语句之前，使用 setXXX(index, value) 方法为 SQL 语句中的参数赋值，其中 index 表示参数的索引，第一个参数的索引为 1，第二个参数的索引为 2，以此类推；value 表示为参数赋予的值；XXX 表示 JDBC 支持的数据类型。然后调用 execute、executeQuery 或 executeUpdate 方法来执行这个 SQL 语句。每一次执行该 SQL 语句前，都可以为参数重新赋值，从而提高执行效率。PreparedStatement 和 Statement 两者的常用方法类似，PreparedStatement 的使用实例如下所示。

```
String sql = "update t_userset name= ?whereid = ?";      // 创建 sql 语句
PreparedStatement pstmt = con.prepareStatement(sql);     // 创建 PreparedStatement 对象
pstmt.setString(1, " 小明 ");                              // 设置第 1 个 sql 参数的值
pstmt.setInt(2, 1);                                       // 设置第 2 个 sql 参数的值
pstmt.executeUpdate();                                    // 执行 sql 语句
```

ResultSet 定义了访问数据库结果集的方法，结果集通常通过执行查询数据库的语句生成。常用的方法如表 3.4 所示，其他方法可查阅 JDK 文档进行学习。

表 3.4 ResultSet 的常用方法

方法名	方法描述
void close()throws SQLException	立即释放此 ResultSet 对象的数据库和 JDBC 资源
boolean absolute(int row) throws SQLException	将光标移动到 ResultSet 对象的指定行，如果 row 为负数，则放到倒数第几行
boolean next()throws SQLException	将光标从当前位置向后移一行（ResultSet 光标最初位于第一行之前）
boolean previous()throws SQLException	将光标移动到当前行的上一行

ResultSet 通过 getXXX 方法获取一条记录中的数据项，其中 XXX 表示 JDBC 中支持的数据类型。同时需要指定需要获取的数据项，有两种方式：一种是用一个 int 值作为数据项的索引，另一种是用列名作为数据项的索引。ResultSet 的使用如下所示。

```
String sql = "select * from t_user";
Statement stmt = con.createStatement();
ResultSet rs = stmt.executeQuery(sql);
while(rs.next()) {
System.out.println(rs.getInt(1)); // 输出当前行第 1 列的值
    System.out.println(rs.getString("name")); // 输出当前行 "name" 列的值
}
```

③ JDBC 连接数据库的步骤

● 导入 JDBC 包

使用的数据库不同，导入的数据库驱动也不同，本书将使用 MySQL 数据库，MySQL 的 JDBC 驱动为 mysql-connector-java-3.1.13-bin.jar。引入 jar 包的方法为：直接

将 jar 包复制（或拖拽）到 JavaWeb 项目中的 \WebContent\WEB-INF\lib 目录下即可。

● 加载数据库驱动

Java 中最常用的加载驱动程序的方式是通过 Class.forName() 完成，它会将驱动程序的类文件加载到内存中，下面的示例将使用 Class.forName() 来加载驱动程序。

```
try {
Class.forName("com.mysql.jdbc.Driver");
}catch (ClassNotFoundException e) {
System.out.println(" 加载数据库驱动失败 ");
e.printStackTrace();
}
```

● 获取数据库连接

加载完驱动程序后，就可以使用 DriverManager.getConnection() 方法获取数据库连接，DriverManager 提供了三个重载的方法，具体如下：

public static Connection getConnection(String url)
public static Connection getConnection(String url, Properties info)
public static Connection getConnection(String url, String user, String password)

在获取数据库连接时需要指定数据库的 URL，表 3.5 将列举常见数据库的 URL 格式和默认端口号。

表 3.5 常见数据库的 URL 格式

数据库	URL 格式	说明
MySQL	jdbc:mysql://hostname:port/databaseName	默认端口为 3306
Oracle	jdbc:oracle:thin:@hostname:port/databaseName	默认端口为 1521
DB2	jdbc:db2://hostname:port/databaseName	默认端口为 6789
Sybase	jdbc:jtds:Sybase://hostname:port/databaseName	默认端口为 2638
SQL Server	jdbc:microsoft:sqlserver://hostname:port;databaseName	默认端口为 1433

本示例通过 DriverManager.getConnection(String url, String user, String password) 方法获取数据库连接，具体代码如下：

```
Connectionconn = null;
try {
Class.forName("com.mysql.jdbc.Driver");
conn = DriverManager.getConnection("jdbc:MySQL://127.0.0.1:3306/info", "root", "root");
}catch (ClassNotFoundException e) {
System.out.println(" 加载驱动失败 ");
e.printStackTrace();
}catch (SQLException e) {
e.printStackTrace();
}
```

● 关闭数据库连接

在数据库程序结束之后，必须关闭所有的连接对象。虽然 Java 虚拟机会关闭连接，但关闭行为不能被程序员控制，关闭时间并不确定，因此在程序中及时关闭连接是较好的编程习惯。

不管程序是否抛出异常，都需要关闭连接，因此需要将关闭操作放到 finally 块中。具体的示例代码如下：

```java
finally {
try {
   if(null !=conn) {
     conn.close();
   }
} catch(SQLException e) {
   e.printStackTrace();
}
}
```

④使用 JDBC 实现数据库的增删改查

● 创建数据库和表

在用 JDBC 操作数据库之前，首先要建立数据库和表，创建数据库和表的具体脚本如下：

```sql
drop database if exists info;    /* 如果数据库存在则删除 */
create database info;            /* 创建数据库 */
use info;                        /* 使用数据库 */
create table t_user(             /* 创建数据库表 */
id INT NOT NULL AUTO_INCREMENT,
name VARCHAR(100) NOT NULL,
password VARCHAR(40) NOT NULL,
sex VARCHAR(40) NOT NULL,
birthday VARCHAR(40) NOT NULL,
interest VARCHAR(40),
PRIMARY KEY (id)
);
```

● 通过 JDBC 操作数据库

对数据库的操作包括插入、删除、修改、查询，具体的代码如下：

```java
Test.java
public class Test {
    private Connection conn;        // 数据库连接对象
    private Statement stmt;         // 声明执行 SQL 语句的容器
private String driver = "com.mysql.jdbc.Driver";
private String url = "jdbc:mysql://localhost:3306/info?characterEncoding=UTF-8";
private String username = "root";
private String password = "root";
   // 获取数据库连接
public boolean connect() {
   try {
      Class.forName(driver);
      conn = DriverManager.getConnection(url,username, password);
      stmt = conn.createStatement();
      System.out.println(" 装载 JDBC 驱动程序成功 ");
      System.out.println(" 数据库连接成功 ");
      return true;
```

```java
    } catch (Exception e) {
    System.out.println(" 装载 JDBC 驱动程序失败 ");
        System.out.println(" 数据库连接失败 ");
    return false;
    }
}
    // 关闭数据库连接
    public void disConnection() {
    try {
      if (null!=conn) {
         conn.close();
        }
      }catch (Exception e) {
    System.out.println(" 关闭数据库失败 ");
}
}

    /**
     * 查询数据
     * @param sql：查询语句
     * @return 查询结果
     */
public ResultSet select(String sql) {
        ResultSet rs = null;
    try {
      rs = stmt.executeQuery(sql);
        } catch (SQLException e) {
    System.out.println("select(): 执行查询语句失败 ");
        }
    return rs;
    }
    /**
     * 插入数据
     * @param sql：插入语句
     * @return true 表示插入成功，false 表示插入失败
     */
public boolean insert(String sql) {
    try {
      stmt.executeUpdate(sql);
       return true;
        } catch (Exception e) {
      System.out.println("insert(): 执行插入语句失败 ");
      return false;
        }
    }
    /**
     * 删除数据
     * @param sql：删除语句
     * @return true 表示删除成功，false 表示删除失败
     */
```

```java
public boolean delete(String sql) {
    try {
        stmt.executeUpdate(sql);
        return true;
    } catch (Exception e) {
        System.out.println("delete(): 执行删除语句失败 ");
        return false;
    }
}
/**
 * 修改数据
 * @param sql：修改语句
 * @return true 表示修改成功，false 表示修改失败
 */
public boolean update(String sql) {
    try {
        stmt.executeUpdate(sql);
        return true;
    } catch (Exception e) {
        System.out.println("update(): 执行修改语句失败 ");
        return false;
    }
}

public static void main(String[] args) {
    String insertSQL = "insert into t_user(name, password,sex,birthday) values('admin','123456','M','1990-8-8')";
    String selectSql = "select * from t_user";
    String deleteSql = "delete from t_user";
    ResultSet rs;
    int i = 0;
    Test test = new Test();
    if(test.connect()) {            // 连接数据库
        if(test.insert(insertSQL)) {   // 执行插入语句
            System.out.println(" 插入成功 ");
            rs = test.select(selectSql); // 执行查询语句
            try {
                while(rs.next()) {
                    System.out.print(" 第 "+ (++i) + " 条记录 :(" + rs.getInt(1) + "," + rs.getString(2) + "," + rs.getString(3) + "," + rs.getString(4) + "," + rs.getString(5) + ")");
                    System.out.println();
                }
            } catch (Exception e) {
                e.printStackTrace();
            }
            test.delete(deleteSql);  // 执行删除语句
            System.out.println(" 删除成功 ");
        } else {
            System.out.println(" 插入失败 ");
        }
```

```
        }
        test.disConnection();           // 关闭连接
    }
}
```
● 调试程序，运行结果

该程序的部分运行结果如图 3.1 所示。

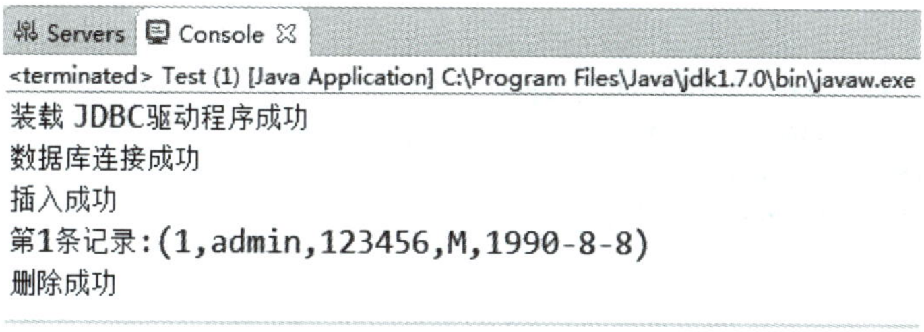

```
装载 JDBC驱动程序成功
数据库连接成功
插入成功
第1条记录:(1,admin,123456,M,1990-8-8)
删除成功
```

图 3.1　数据库操作部分运行结果

3.1.2　JDBC 的封装

通过 3.1.1 节的学习可知，在对数据库操作之前，先要加载驱动程序，获取数据库连接，再获取 Statement 或 PreparedStatement 对象实现对数据库的操作，操作完成后要关闭数据库资源。这些步骤都是重复性的，如果每次操作都重复这些步骤，会导致程序中出现大量重复代码，不利于维护。为了解决这一问题，本节将对常用的数据库操作进行封装，命名为 DBHelper 类。

封装之后，直接调用 DBHelper.ExecSql 方法执行添加、删除、修改命令，调用 DBHelper. getResultSet 方法执行查询命令并返回结果集。具体代码如下：

```java
DBHelper.java
package DBHelper;
import java.sql.*;
public class DBHelper
{
    // 驱动
    public static String driver = "com.mysql.jdbc.Driver";
    // 连接字符串
    public static String url = "jdbc:mysql://localhost:3306/info?characterEncoding=UTF-8";
    // 用户名
    public static String user = "root";
    // 密码
    public static String password =  "root";

    // 不允许实例化该类
    private DBHelper(){ }

    /**
     * 获取一个数据库连接
     * 通过 driver、url、user、password 这 4 个静态成员来设置数据库连接属性
```

```java
 * @return 数据库连接
 */
public static Connection getConnection() {
    try{
        Class.forName(driver); // 加载驱动
    } catch (ClassNotFoundException ex){
        System.out.println(ex.getMessage());
    }
    try {
        if(user==null) return DriverManager.getConnection(url);
        return DriverManager.getConnection(url, user, password);
    } catch (SQLException ex) {
        System.out.println(ex.getMessage());
        return null;
    }
}
/**
 * 获取一个 Statement
 * 该 Statement 已经设置数据集，可以滚动，可以更新
 * @return 如果获取失败，返回 null
 */
public static Statement getStatement(){
    Connection conn = getConnection();
    if (conn == null) return null;
    try{
        return conn.createStatement(ResultSet.TYPE_SCROLL_SENSITIVE,
            ResultSet.CONCUR_UPDATABLE);
    } catch (SQLException ex) {
        System.out.println(ex.getMessage());
        close(conn);
    }
    return null;
}
/**
 * 获取一个带参数的 PreparedStatement
 * 该 PreparedStatement 已经设置数据集，可以滚动，可以更新
 * @param cmdText 带 "?" 参数的 SQL 语句
 * @param cmdParams SQL 语句的参数表
 * @return 如果获取失败，返回 null
 */
public static PreparedStatement getPreparedStatement(String cmdText, Object[] cmdParams) {
    Connection conn = getConnection();
    if (conn == null) return null;
    PreparedStatement pstmt = null;
    try{
        pstmt = conn.prepareStatement(cmdText, ResultSet.TYPE_SCROLL_SENSITIVE, ResultSet.CONCUR_UPDATABLE);
        int i = 1;
        for (Object item : cmdParams){ // 遍历集合
            pstmt.setObject(i, item);
```

```java
        i++;
      }
    } catch (SQLException e) {
      e.printStackTrace();
      close(conn);
    }
    return pstmt;
}

/**
 * 执行 SQL 语句，返回结果为整型
 * 主要用于执行非查询语句
 * @param cmdText SQL 语句
 * @return：非负数，正常执行；-1，执行错误；-2，连接错误
 */
public static int ExecSql(String cmdText) {
    Statement stmt = getStatement();
    if (stmt == null) return -2;
    int i;
    try {
        i = stmt.executeUpdate(cmdText);
    } catch (SQLException ex) {
            System.out.println(ex.getMessage());
        i = -1;
    }
    closeConnection(stmt);
    return i;
}
/**
 * 执行 SQL 语句，返回结果为整型
 * 主要用于执行非查询语句
 * @param cmdText 带 "?" 参数的 SQL 语句
 * @param cmdParams SQL 语句的参数表
 * @return：非负数，正常执行；-1，执行错误；-2，连接错误
 */
public static int ExecSql(String cmdText, Object[] cmdParams) {
    PreparedStatement pstmt = getPreparedStatement(cmdText, cmdParams);
    if (pstmt == null) return -2;
    int i;
    try {
        i = pstmt.executeUpdate();
    } catch (SQLException ex) {
      System.out.println(ex.getMessage());
      i = -1;
    }
    closeConnection(pstmt);
    return i;
}

/**
```

```java
 * 返回一个 ResultSet
 * @param cmdText SQL 语句
 * @return
 */
public static ResultSet getResultSet(String cmdText) {
    Statement stmt = getStatement();
    if (stmt == null) return null;
    try {
        return stmt.executeQuery(cmdText);
    } catch (SQLException ex) {
            System.out.println(ex.getMessage());
        closeConnection(stmt);
    }
    return null;
}
/**
 * 返回一个 ResultSet
 * @param cmdText 需要"?"参数的 SQL 语句
 * @param cmdParams SQL 语句的参数表
 * @return
 */
public static ResultSet getResultSet(String cmdText, Object[] cmdParams) {
    PreparedStatement pstmt = getPreparedStatement(cmdText, cmdParams);
    if (pstmt == null) return null;
    try {
        return pstmt.executeQuery();
    } catch (SQLException ex) {
        System.out.println(ex.getMessage());
        closeConnection(pstmt);
    }
    return null;
}

private static void close(Object obj) {
    if (obj == null) return;
    try {
        if (obj instanceof Statement) {
            ((Statement) obj).close();
        } else if (obj instanceof PreparedStatement) {
            ((PreparedStatement) obj).close();
        } else if (obj instanceof ResultSet) {
            ((ResultSet) obj).close();
        } else if (obj instanceof Connection) {
            ((Connection) obj).close();
        }
    } catch (SQLException ex) {
        System.out.println(ex.getMessage());
    }
}
/**
```

```
     * 获取结果集中第一行第一列的值
     * @param ResultSet
     * @return 结果集中第一行第一列的值
     */
    public static Object getScalar(ResultSet rs) {
       if (rs == null) return null;
       Object obj = null;
       try {
          if (rs.next()) obj = rs.getObject(1);
       } catch (SQLException ex) {
                System.out.println(ex.getMessage());
       }
       return obj;
    }
    /**
     * 关闭数据库连接
     * @param 数据库操作相关对象（如 Statement、ResultSet 等）
     */
    private static void closeConnection(Object obj) {
       if (obj == null) return;
       try {
         if (obj instanceof Statement) {
            ((Statement) obj).getConnection().close();
         } else if (obj instanceof PreparedStatement) {
            ((PreparedStatement) obj).getConnection().close();
         } else if (obj instanceof ResultSet) {
            ((ResultSet) obj).getStatement().getConnection().close();
         } else if (obj instanceof Connection) {
            ((Connection) obj).close();
         }
       } catch (SQLException ex) {
         System.out.println(ex.getMessage());
       }
    }
}
```

3.1.3　数据库操作分层设计

数据库操作通常包括业务逻辑代码和数据持久化代码。其中，业务逻辑代码表示处理数据逻辑的代码，数据持久化代码则是实现数据永久保存的代码，持久化的主要方式包括将数据保存到数据库、普通文件和 XML 文件中，主要涉及到的操作有插入、删除、查询和修改。如果将业务逻辑代码和数据持久化代码放到一个文件内，会存在逻辑不够清晰的问题，也不便于代码维护，因此要将两者进行分离。

① DAO 模式

● DAO 模式概述

DAO（Data Access Object）表示数据访问对象，是一个数据访问的接口，位于业务逻辑层与数据库资源访问层之间。DAO 模式会将数据库的所有操作抽象封装在接口中，该接口中定义应用程序中用到的所有数据访问方法，再定义一个实现类实现该接口，从而为业务逻辑层提供支持。

- DAO 模式优势

DAO 模式隔离了业务逻辑代码和数据库访问代码，提高了代码可复用性。当业务逻辑发生改变后，只需要改变业务逻辑层的代码即可，数据库访问层的代码不需要改变。当数据库发生变化后，只需要增加数据库访问接口的实现类即可，降低了耦合性，提高了代码的可扩展性和系统的可移植性。

②分层开发的步骤

DAO 模式的类图如图 3.2 所示。

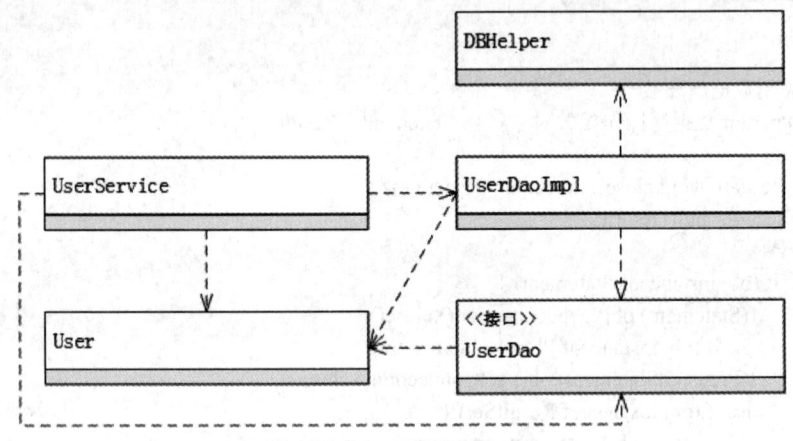

图 3.2　DAO 模式的类图

图 3.2 列举出了数据库分层操作示例中涉及的类和接口之间的关系，UserService、UserDao、UserDaoImpl 三者都会用到 User 类，而 UserDaoImpl 实现 UserDao 接口，以便 UserService 类能够调用，而 UserDaoImpl 类依赖 DBHelper 类，通过调用其中的方法实现数据访问。

分层开发的具体步骤如下所示。

- 创建实体类

在日常 Java 开发中，实体类（entity）是必不可少的。实体类主要封装了多个私有属性并提供相应的 getter 和 setter 方法，以供程序访问，它也包含构造函数，可根据具体情况进行定义。实体类是作为数据传输而存在的，比如作为函数的输入参数或返回类型，一般其包含的属性是和数据库表中的字段相对应的。实体类的定义一般遵循如下规范。

实体类中的属性一般定义为私有属性，即用 private 关键字修饰。

为实体类中的私有属性提供对应的 getter 和 setter 方法，在 Eclipse 中打开类文件，在右键菜单中选择"Source"→"Generate Getters and Setters…"，可以自动生成属性的 getter 和 setter 方法。

在实体类中定义无参数构造器，也可根据设计需要定义带参构造函数。

实体类可以实现 java.io.Serializable 接口，实现了该接口的实体类支持序列化机制，可以将实体类对象以字节的形式保存在磁盘上，当需要时再加载该对象，如果实现了 Serializable 接口，则需要提供一个版本号，用以解决不同版本之间的序列化问题。下面将以 User 类为例演示如何定义一个实体类。

```java
User.java
package com.entity;
public class User {
    private int id;
    private String name;
    private String password;
    private String sex;
    private String birthday;
    private String interest;

    public User() {}
    public User(String name, String password, String sex, String birthday, String interest) {
        this.name = name;
        this.password = password;
        this.sex = sex;
        this.birthday = birthday;
        this.interest = interest;
    }

    /* 各属性的 getter 和 setter 方法代码略 */
}
```

- 创建 DAO 层

DAO 层主要是封装数据库访问接口，为 Service 层提供数据访问方法，DAO 层主要包括 DBHelper 类、Dao 接口和 Dao 接口的实现类，其中 DBHelper 类为 DAO 层提供统一的插入、删除、修改和查询等操作，避免重复代码的出现，DBHelper 的具体实现如 3.1.2 节所示。

```java
//UserDao 接口类（UserDao.java）
package com.dao;
public interface UserDao {
    int insert(User user);
    int delete(User user);
    int update(User user);
    List<User>select(User user);
    User selectById(String id);
    boolean isExist(User user);
}

// UserDao 接口的实现类（UserDaoImpl.java）
package com.dao.impl;
import java.util.List;
import com.dao.UserDao;
import com.entity.User;
import com.utill.DBHelper;
```

```java
public class UserDaoImpl implements UserDao {

    // 将 user 插入数据库
    @Override
    public int insert(User user) {
        String sql = "insert into t_user(name, password, sex, birthday, interest) values(?, ?, ?, ?, ?)";
        Object[] objs = {user.getName(), user.getPassword(), user.getSex(), user.getBirthday(), user.getInterest()};
        return DBHelper.ExecSql (sql, objs);
    }
    // 将 user 从数据库中删除
    @Override
    public int delete(User user) {
        String sql = "delete from t_user where id = ?";
        return DBHelper.ExecSql (sql, new Object[]{user.getId()});
    }

    // 修改 user 对象，并持久化到数据库
    @Override
    public int update(User user) {
        return 0;
    }

    // 查询 user 对象
    @Override
    public List<User> select(User user) {
        return null;
    }

    // 查询 id 对应的数据库记录
    @Override
    public User selectById(String id) {
        return null;
    }

    // 查询 user 对象是否存在于数据库中
    @Override
    public boolean isExist(User user) {

    }
}
```

● 创建 Service 层

Service 层主要编写业务逻辑代码，本任务中主要完成注册功能，在向数据库中插入用户信息之前，要先验证该用户名是否已经存在，如果已经存在，应提示用户。

UserService.java
```java
package com.service;
public class UserService {
    private UserDao userDao = new UserDaoImpl();
    /**
```

```
 * 用户注册
 * @param user：用户注册信息
 * @return：返回 -1 表示用户名重复，0 表示添加失败，正数表示影响的行数
 */
public int register(User user) {
    if(userDao.isExist(user)) { // 表示该用户名已经存在
        return -1;
    }
    return userDao.insert(user);
}
```

【任务实施】

1. 创建数据库和表

首先需要创建数据库和表，具体代码如下所示，若库已经存在数据库 info，会先将其删除，否则直接创建数据库 info 和表 t_user。

```sql
drop database if exists info;
create database info;
use info;
create table t_user(
id INT NOT NULL AUTO_INCREMENT,
name VARCHAR(100) NOT NULL,
password VARCHAR(40) NOT NULL,
sex VARCHAR(40) NOT NULL,
birthday VARCHAR(40) NOT NULL,
interest VARCHAR(40),
PRIMARY KEY (id)
);
```

2. 创建 DBHelper 类

为了简化对数据库表的操作，避免重复代码的出现，需要对数据库的操作进行封装并命名为 DBHelper 类，具体代码参考相关知识中的"JDBC 封装"。

3. 创建实体类、Dao 接口及实现类

用户的实体（User）类对用户的属性进行封装，业务逻辑层将用户信息以 user 对象的形式传递给数据访问层，User 类代码如下：

```java
User.java
package com.entity; // 包名一般带有 entity、bean、po 等字样，表示包含实体类
public class User {
    private int id;
    private String name;
    private String password;
    private String sex;
    private String birthday;
```

```java
    private String interest;
    public User() {}
    public User(String name, String password, String sex, String birthday, String interest) {
        this.name = name;
        this.password = password;
        this.sex = sex;
        this.birthday = birthday;
        this.interest = interest;
    }
    /* 各属性的 getter 和 setter 方法代码略 */
}
```

数据访问层接口及实现类代码如下：

```java
UserDao.java
package com.dao;
public interface UserDao {
    int insert(User user);
    int delete(User user);
    int update(User user);
    List<User> select(User user);
    User selectById(String id);
    boolean isExist(User user);
}
// UserDao 接口的实现类 (UserDaoImpl.java)
package com.dao.impl;
public class UserDaoImpl implements UserDao {
    // 将 user 插入数据库
    @Override
    public int insert(User user) {
        String sql = "insert into t_user(name, password, sex, birthday, interest) values(?, ?, ?, ?, ?)";
        Object[] objs = {user.getName(), user.getPassword(), user.getSex(), user.getBirthday(), user.getInterest()};
        return DBHelper.ExecSql(sql, objs);
    }
    // 将 user 从数据库中删除
    @Override
    public int delete(User user) {
        String sql = "delete from t_user where id = ?";
        return DBHelper.ExecSql(sql, new Object[]{user.getId()});
    }
    // 修改 user 对象，并持久化到数据库（尚未实现）
    @Override
    public int update(User user) {
        return 0;
    }
    // 查询 user 对象（尚未实现）
    @Override
    public List<User> select(User user) {
        return null;
    }
```

```java
// 查询 id 对应的数据库记录（尚未实现）
    @Override
    public User selectById(String id) {
        return null;
    }
// 查询 user 对象是否存在于数据库中（尚未实现）
    @Override
    public boolean isExist(User user) {
        return false;
    }
}
```

4. 创建 Service 类

UserService 类对外提供用户管理的相关服务，目前只实现了用户注册功能，类的具体代码如下：

UserService.java
```java
package com.service;
public class UserService {
    private UserDao userDao = new UserDaoImpl();
    /**
     * 用户注册
     * @param user：用户注册信息
     * @return：返回 -1 表示用户名重复，0 表示添加失败，正数表示影响的行数
     */
    public int register(User user) {
        if(userDao.isExist(user)) { // 表示该用户名已经存在
            return -1;
        }
        return userDao.insert(user);
    }
}
```

5. 创建测试类

接下来测试 UserService 类的 register 方法功能是否正常，代码如下：

Test3.java
```java
public class Test3 {
    public static void main(String[] args) {
        UserService service = new UserService();
        User user = new User(" 小明 ", "123456","M","1990-8-8");
        int result = service.register(user);
        if(-1 == result) {
            System.out.println(" 用户名已经存在 ");
        } else if(0 == result){
            System.out.println(" 注册失败 ");
        } else {
            System.out.println(" 注册成功 ");
        }
    }
}
```

第一次程序运行的结果如图 3.3 所示。

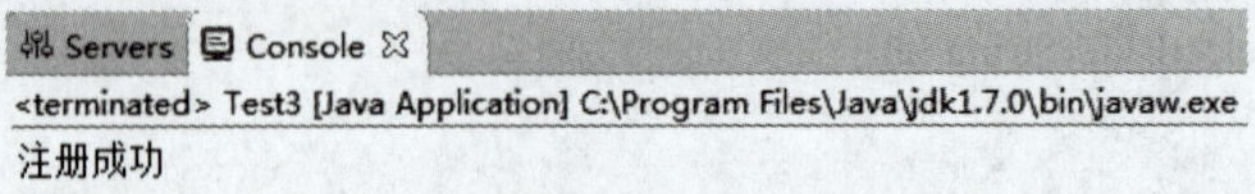

```
Servers  Console ⊠
<terminated> Test3 [Java Application] C:\Program Files\Java\jdk1.7.0\bin\javaw.exe
注册成功
```

图 3.3　数据库分层操作测试结果一

第二次程序运行的结果如图 3.4 所示。

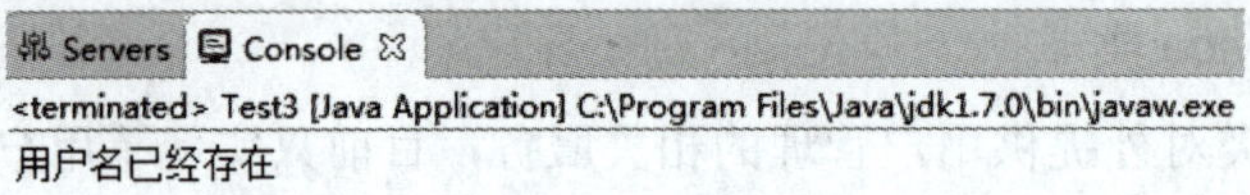

```
Servers  Console ⊠
<terminated> Test3 [Java Application] C:\Program Files\Java\jdk1.7.0\bin\javaw.exe
用户名已经存在
```

图 3.4　数据库分层操作测试结果二

【实践训练】

（1）简述连接数据库的步骤。
（2）简述数据库操作分为几层，各完成什么功能，以及数据库分层操作的优势。
（3）完成 3.1.3 节中 UserDaoImpl 的其他功能，如更新和查询等。

任务 3.2　用 Servlet 接收和处理数据

【任务分析】

本任务设计用户注册页面并验证用户输入数据的合法性。先用 Servlet 接收用户注册的信息，再验证输入数据的合法性以及数据库中用户名是否存在：若数据通过检验，则保存用户信息到数据库并跳转到用户注册成功提示页面；否则跳转到错误信息提示页面。

本任务采用分层方式进行设计，Servlet 收到客户端提交的数据后，将用户信息存入实体类对象中，再传递给 Service 层（业务逻辑层）进行处理。

【相关知识】

3.2.1　Servlet 基础

访问 Web 应用程序的过程就是请求和响应的过程。用户在浏览器中输入网址并按回车键后,浏览器会向服务器发送一个 HTTP 请求,服务器接收到请求后会做相应的处理,并为用户做出恰当的响应。该访问过程是基于 HTTP 的,在 2.1.1 节中已经介绍过 HTTP,HTTP 提供了 GET、POST、HEAD、DELETE、TRACE、PUT 和 OPTIONS 等 7 种访问方式,其中最常用的是 GET 和 POST 这两种方式。

① Servlet 概述

Servlet 是 Java Web 应用程序的核心,它能够处理请求和并给出响应,Servlet 中定义了很多方法,开发人员仅需要实现恰当的方法来响应用户请求即可。Servlet 寄生在 Tomcat 服务器这个容器中,服务器在固定端口(如 8080)监听,当收到请求后,会自动调用 Servlet 的相应方法(如 doGet、doPost)处理收到的数据,完成响应过程。

- Servlet 工作流程

用户使用浏览器提交一个请求,该请求遵循 HTTP,Tomcat 接收到请求后,对请求进行解析,并封装成 HttpServletRequest 类型的 request 对象,可通过此对象获得 HTTP 头数据。Tomcat 也会把输出流封装成 HttpServletResponse 类型的 response 对象,可通过此对象输出响应内容。然后 Tomcat 会将这两个对象作为输入参数调用 Servlet 的相应方法,比如 doGet(request, response) 或 doPost(request, response) 方法等,我们就可以在调用的方法中实现程序的业务逻辑。Servlet 的典型请求响应过程如图 3.5 所示。

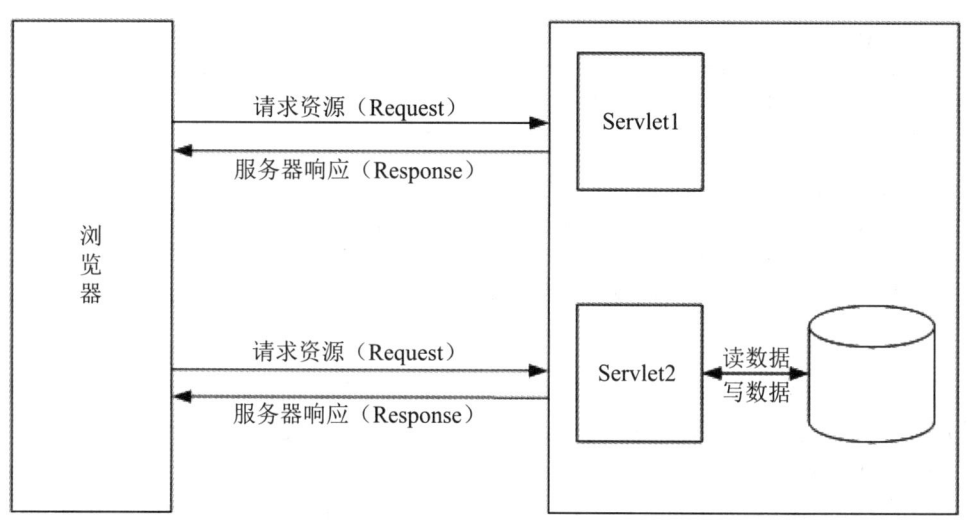

图 3.5　Servlet 请求响应过程

- Servlet 接口与实现类

Servlet 接口位于包 javax.servlet 内,所有 Servlet 都需要直接实现这一接口或者继承实现了该接口的类。该接口有两个实现类:GenericServlet 和 HttpServlet。这两个类均为抽象类,大多数情况下,开发人员只需要在这两个类的基础上进行扩展即可实现各自的功能,其中最常用的是 HttpServlet,HttpServlet 类位于 javax.servlet.http 包内。HttpServlet 类中常用的方法如表 3.6 所示,其他方法可查阅 Java EE 文档,官方网址是:http://docs.oracle.com/javaee/6/api/。

表 3.6　HttpServlet 类的常用方法

方法名	方法描述
protected void doGet(HttpServletRequest req, HttpServletResponse res) throws ServletException, IOException	当浏览器以 GET 方式访问时被调用
protected void doPost(HttpServletRequest req, HttpServletResponse res) throws ServletException, IOException	当浏览器以 POST 方式访问时被调用
protected long getLastModified(HttpServletRequest req)	返回 HttpServletRequest 最后修改的时间

②编写 Servlet

直接实现 Servlet 接口来编写 Servlet 需要实现很多方法，而 HttpServlet 是 Servlet 接口的一个实现类，一般情况下，编写的 Servlet 类需要继承 HttpServlet 类，并覆盖其中的方法，通常只覆盖 doGet 方法和 doPost 方法。

● 实现 Servlet 类

开发人员只需要定义一个继承 HttpServlet 类的普通 Java 类，并覆盖 doGet 方法和 doPost 方法，一个能够处理请求和响应的 Servlet 类就定义好了，具体的代码如下：

```
UserServlet.java
public class UserServletextends HttpServlet {
  @Override
protected void doGet(HttpServletRequest request, HttpServletResponse response)
throws ServletException, IOException {
  System.out.println("UserServlet：doGet 方法被调用 ");
  }
  @Override
protected void doPost(HttpServletRequest req, HttpServletResponseresp)
throws ServletException, IOException {
    System.out.println("UserServlet：doPost 方法被调用 ");
  }
}
```

当客户端浏览器以 GET 方式发送请求时，控制台中会打印"UserServlet：doGet 方法被调用"；以 POST 方式发送请求，控制台中会打印"UserServlet：doPost 方法被调用"。

● 配置 Servlet

在定义好 Servlet 类文件之后，还要通过配置告知 Web 容器该 Servlet 的类名、参数、访问路径等信息，配置信息存放在项目的 /WebContent/WEB-INF/web.xml 中，具体的配置内容如下：

```xml
<servlet>
  <servlet-name>UserServlet</servlet-name>
  <servlet-class>com.servlet.UserServlet</servlet-class>
  <init-param>
    <param-name>name</param-name>
    <param-value>tom</param-value>
  </init-param>
  <init-param>
    <param-name>encoding</param-name>
```

```
        <param-value>utf-8</param-value>
    </init-param>
    <load-on-startup>1</load-on-startup>
</servlet>
```

<servlet> 与 </servlet> 分别表示开始标签和结束标签，中间内容则为 Servlet 的具体配置信息。

<servlet-name> 与 </servlet-name> 标签是必须配置的，代表 Servlet 的名字，名字可以任意指定，但须保证在 web.xml 文件中的唯一性。

<servlet-class> 与 </servlet-class> 标签也是必须配置的，代表 Servlet 类名，由于类名较长，容易书写错误，因此在写好类名后，可以在按住 Ctrl 键的同时，用鼠标左键点击类名，若能够跳转到类中，证明书写正确，否则书写错误。

<init-param> 与 </init-param> 标签是可选的，用于配置初始化参数，该标签包含两个子标签，<param-name> 与 </param-name> 标签用于指定参数名称，<param-value> 与 </param-value> 标签用于指定参数值。在 Servlet 类中可以通过 getServletConfig().getInitParameter(paramName) 来获取初始化参数，<init-param> 标签可以配置多个。

<load-on-startup> 与 </load-on-startup> 标签用于<u>标记是否在容器启动时就加载这个 servlet</u>，当值为 0 或者大于 0 时，表示容器在应用启动时就加载这个 Servlet；当是一个负数或者没有指定时，则指示容器在该 Servlet 被选择时才加载。正数的值越小，启动该 Servlet 的优先级越高。

● 配置 <servlet-mapping>

Web 容器不仅需要知道类文件的位置，也需要知道此 Servlet 对应的路径（URL 地址）。该配置信息须放在 web.xml 中，具体的配置代码如下：

```
<servlet-mapping>
    <servlet-name>UserServlet</servlet-name>
    <url-pattern>/UserServlet</url-pattern>
    <url-pattern>/UserServlet.jsp</url-pattern>
    <url-pattern>/UserServlet.php</url-pattern>
</servlet-mapping>
```

<servlet-mapping> 与 </servlet-mapping> 分别表示开始标签和结束标签，中间内容则为 servlet-mapping 的配置信息。

<servlet-name> 与 </servlet-name> 标签是必须配置的，代表 Servlet 的名字，与前面 <servlet> 配置信息中的 <servlet-name> 一致。

<url-pattern> 与 </url-pattern> 标签是必须配置的，代表此 Servlet 能够处理的 URL 路径，一个 <servlet-mapping> 标签可以配置多个 <url-pattern> 标签。用户通过浏览器访问任何一个已经配置的 <url-pattern> 中的 URL，Tomcat 都会自动跳转到对应的 Servlet 内执行 doXXX 方法，XXX 取决于用户的访问方式，若以 Get 方式访问，则执行 doGet 方法，若以 Post 方式访问，则执行 doPost 方法。<url-pattern> 前面加上服务器域名和端口号即为该 Servlet 的访问网址，本例中的 Servlet 访问网址如下：

http://localhost:8080/servlet/UserServlet
http://localhost:8080/servlet/UserServlet.jsp

http://localhost:8080/servlet/UserServlet.php

访问这三个网址中的任意一个，都会跳转到 UserServlet 中，从而实现隐藏编程语言的目的，用户无法从 URL 判断程序代码是用 Java 还是 PHP 编写的。

完整的 web.xml 内容如下：

web.xml

```xml
<?xml version="1.0" encoding="UTF-8"?>
<web-app xmlns:xsi="http://www.w3.org/2001/XMLSchema-instance"
  xmlns="http://java.sun.com/xml/ns/javaee" xsi:schemaLocation="http://java.sun.com/xml/ns/javaee
  http://java.sun.com/xml/ns/javaee/web-app_2_5.xsd" id="WebApp_ID" version="2.5">
  <servlet>
    <servlet-name>UserServlet</servlet-name>
    <servlet-class>com.servlet.UserServlet</servlet-class>
    <init-param>
      <param-name>name</param-name>
      <param-value>tom</param-value>
    </init-param>
    <init-param>
      <param-name>encoding</param-name>
      <param-value>utf-8</param-value>
    </init-param>
    <load-on-startup>1</load-on-startup>
  </servlet>
  <servlet-mapping>
    <servlet-name>UserServlet</servlet-name>
    <url-pattern>/UserServlet</url-pattern>
    <url-pattern>/UserServlet.jsp</url-pattern>
    <url-pattern>/UserServlet.php</url-pattern>
  </servlet-mapping>
</web-app>
```

一个完整的 Servlet 由 Servlet 类、Servlet 配置组成，两者缺一不可，编写完成之后，就可以部署 Web 程序了。

● 部署 Web 程序

在项目 1 已经学习了如何利用 Eclipse 部署 Web 程序，部署完毕后启动 Tomcat 服务器，分别在浏览器输入配置好的 3 个 URL，运行结果如图 3.6 所示。

```
Servers   Console ⊠
Tomcat v7.0 Server at localhost [Apache Tomcat] C:\Program Files\Java\jdk1.7.0\bin\javaw.exe
UserServlet: doGet方法启动
UserServlet: doGet方法启动
UserServlet: doGet方法启动
```

图 3.6 Servlet 运行结果

③处理请求与响应

用户通过浏览器向 Tomcat 服务器发送请求，Tomcat 会根据配置信息分配给相应 Servlet 的方法（如 doGet、doPost）进行处理，一般传递给方法的参数有两个，参数类别分别为 HttpServletRequest 和 HttpServletResponse。例如在 UserServlet.java 中，doGet

方法定义为：

protected void doGet(HttpServletRequest request, HttpServletResponse response)

● 请求（HttpServletRequest）

浏览器发送的请求被封装成为一个 HttpServletRequest 类型的对象，HttpServlet-Request 接口位于包 javax.servlet.http 内，通过该对象可以获取请求的地址、请求的参数、提交的数据、上传的文件、客户端的 IP 地址甚至是客户端操作系统等信息。HttpServletRequest 中的常用方法如表 3.7 所示，其他方法可查阅 Java EE 文档。

表 3.7　HttpServletRequest 的常用方法

方法名	方法描述
Object getAttribute(String name)	获取名为 name 的属性值，若不存在则返回 null
void setAttribute(String name, Object o)	向 request 中存储属性 name，值为对象 o
void removeAttribute(String name)	从 request 中移除属性 name
String getParameter(String name)	获取名为 name 的请求参数值，若不存在则返回 null
String[] getParameterValues(String name)	获取名为 name 的属性值数组，若不存在则返回 null
RequestDispatcher getRequestDispatcher(String path)	获取路径 path 对应的 RequestDispatcher 对象，用于向一个资源转发请求
void setCharacterEncoding(Stringenv)throws UnsupportedEncodingException	设置 request 的编码格式
Cookie[] getCookies()	获取客户端发送的 Cookie 数组

● 响应（HttpServletResponse）

服务器对浏览器的响应被封装成为一个 HttpServletResponse 类型的对象，HttpServlet-Response 接口位于包 javax.servlet.http 内，HttpServletResponse 中的常用方法如表 3.8 所示，其他方法可查阅 Java EE 文档。

表 3.8　HttpServletResponse 的常用方法

方法名	方法描述
ServletOutputStream getOutputStream()throws IOException	获取 ServletOutputStream 对象，用于响应二进制数据
PrintWriter getWriter()throws IOException	获取 PrintWriter 对象，用于响应字符数据
void sendRedirect(String location) throws IOException	向客户端发送一个临时重定向响应消息
void setCharacterEncoding(String charset)	设置 response 的编码格式
void setContentType(String type)	设置 response 的内容类型
void addCookie(Cookie cookie)	向 response 中添加 cookie 对象

下面代码展示了在 Servlet 中使用 request 获取客户端的信息并通过 response 响应客户端的请求。

```java
@Override
protected void doGet(HttpServletRequest request, HttpServletResponse response) throws ServletException, IOException {
    String requestUrl = request.getRequestURL().toString();        // 请求的 URL 地址
    String requestUri = request.getRequestURI();                    // 请求的资源
    String queryString = request.getQueryString();                  // 请求的 URL 地址中带的参数
    String remoteAddr = request.getRemoteAddr();                    // 来访者的 IP 地址
    String remoteHost = request.getRemoteHost();                    // 来访者的主机名称
    int remotePort = request.getRemotePort();                       // 来访者使用的端口
    String method = request.getMethod();                            // 请求 URL 时用的方法
    String localAddr = request.getLocalAddr();                      //Web 服务器的 IP 地址
    String localName = request.getLocalName();                      //Web 服务器的主机名称
    response.setHeader("content-type", "text/html;charset=UTF-8");  // 设置响应头
    PrintWriter out = response.getWriter();
    out.write(" 获取到的客户机信息如下：<br>");
    out.write(" 请求的 URL 地址："+requestUrl+"<br>");
    out.write(" 请求的资源："+requestUri+"<br>");
    out.write(" 请求的 URL 地址中附带的参数："+queryString+"<br>");
    out.write(" 来访者的 IP 地址："+remoteAddr+"<br>");
    out.write(" 来访者的主机名："+remoteHost+"<br>");
    out.write(" 使用的端口号："+remotePort+"<br>");
    out.write(" 请求使用的方法："+method+"<br>");
    out.write("localAddr: "+localAddr+"<br>");
    out.write("localName: "+localName);
}
```

在浏览器中输入 http://127.0.0.1:8080/servlet/UserServlet.php?name=tom 并按回车键，运行的结果如图 3.7 所示。

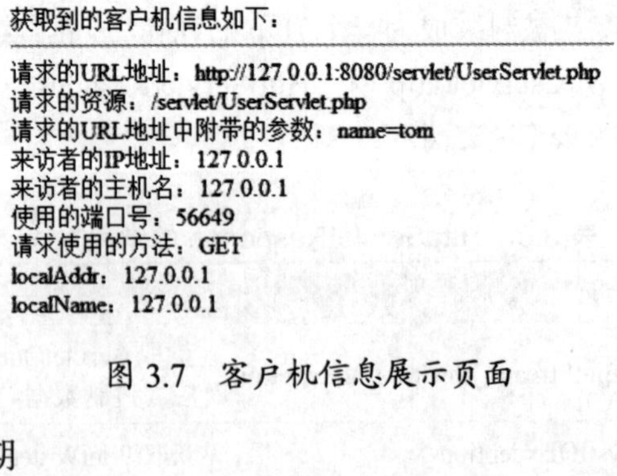

```
获取到的客户机信息如下：
请求的URL地址： http://127.0.0.1:8080/servlet/UserServlet.php
请求的资源： /servlet/UserServlet.php
请求的URL地址中附带的参数： name=tom
来访者的IP地址： 127.0.0.1
来访者的主机名： 127.0.0.1
使用的端口号： 56649
请求使用的方法： GET
localAddr: 127.0.0.1
localName: 127.0.0.1
```

图 3.7 客户机信息展示页面

④ Servlet 生命周期

当服务器启动时，会检查 web.xml 中配置的 Servlet 的 <load-on-startup> 标签的值，若没有配置或者为负数，那么服务器会在第一次使用时才初始化一个 Servlet 对象，否则会在服务器启动时就初始化一个 Servlet 对象，然后服务器会用这个 Servlet 对象处理所有客户端的请求，无论请求多少次，最多只存在一个 Servlet 实例，当多个客户端并发请求 Servlet 时，服务器会启动多个线程分别执行该 Servlet 的 service() 方法。

Servlet 生命周期包括三个阶段，分别执行三个方法，即 init(ServletConfig conf)、service(ServletRequest request, ServletResponse response) 和 destroy()。

- init(ServletConfig conf)

当 <load-on-startup> 标签的值没有配置或者为负数时，则在第一次使用此 Servlet 时才执行此方法，否则服务器启动时就会执行此方法，HttpServlet 的实现类提供了一个更简单的不带参数的 init() 方法，该方法在 Servlet 生命周期中只会执行一次，因此可以将初始化资源的代码放在该函数中执行。

- service(ServletRequest request, ServletResponse response)

HttpServlet 实现类提供了 service(HttpServletRequest request, HttpServletResponse response) 方法。客户端每次请求 Servlet 时，都会运行 service 方法，service 方法会根据访问类型决定需要执行的方法，如 doGet 方法、doPost 方法、doPut 方法。

- destroy()

当容器关闭时，会先卸载所有的 Servlet，卸载 Servlet 时就会执行 destroy() 方法，该方法在 Servlet 生命周期中只会执行一次，因此可以将销毁资源的代码放在该函数中执行。

下面将通过继承 HttpServlet 类并覆盖生命周期的方法来体会 Servlet 的生命周期，具体代码如下。

```java
UserServlet.java
package com.servlet;
public class UserServlet extends HttpServlet {
    @Override
    public void init() throws ServletException {
        System.out.println("init() 方法开始执行 ");
    }
    @Override
    protected void service(HttpServletRequest request, HttpServletResponse response) throws ServletException, IOException {
        System.out.println("service() 方法开始执行 ");
        super.service(request, response);
    }

    @Override
    protected void doGet(HttpServletRequest request, HttpServletResponse response) throws ServletException, IOException {
        System.out.println("doGet() 方法开始执行 ");
    }

    @Override
    protected void doPost(HttpServletRequest req, HttpServletResponse resp) throws ServletException, IOException {
        System.out.println("doPost() 方法开始执行 ");
    }

    @Override
```

```
public void destroy() {
    System.out.println("destroy() 方法开始执行 ");
}
```
}

启动 Tomcat 服务器并在浏览器中访问 http://127.0.0.1:8080/servlet/UserServlet.php 后，关闭 Tomcat 服务器，在控制台会打印如图 3.8 所示的信息。

```
INFO: Starting Servlet Engine: Apache Tomcat/7.0.52
init()方法开始执行
2016-9-17 20:22:36 org.apache.coyote.AbstractProtocol start
INFO: Starting ProtocolHandler ["http-bio-8080"]
2016-9-17 20:22:36 org.apache.coyote.AbstractProtocol start
INFO: Starting ProtocolHandler ["ajp-bio-8009"]
2016-9-17 20:22:36 org.apache.catalina.startup.Catalina start
INFO: Server startup in 908 ms
service()方法开始执行
doGet()方法开始执行
2016-9-17 20:25:16 org.apache.catalina.core.StandardServer await
INFO: A valid shutdown command was received via the shutdown port. Stopping
2016-9-17 20:25:16 org.apache.coyote.AbstractProtocol pause
INFO: Pausing ProtocolHandler ["http-bio-8080"]
2016-9-17 20:25:16 org.apache.coyote.AbstractProtocol pause
INFO: Pausing ProtocolHandler ["ajp-bio-8009"]
2016-9-17 20:25:16 org.apache.catalina.core.StandardService stopInternal
INFO: Stopping service Catalina
destroy()方法开始执行
2016-9-17 20:25:16 org.apache.coyote.AbstractProtocol stop
INFO: Stopping ProtocolHandler ["http-bio-8080"]
2016-9-17 20:25:16 org.apache.coyote.AbstractProtocol stop
```

图 3.8 Servlet 生命周期测试结果

⑤ Servlet 之间的跳转

Servlet 之间可以相互跳转，通过跳转可以实现将一项任务进行分解，比如用一个 Servlet 专门接收用户的请求和响应，获取用户提交的参数并为用户产生恰当的响应，然后跳转到另外一个 Servlet 专门处理业务逻辑，比如操作数据库，最后再跳转到一个 Servlet 专门为用户显示处理结果。Servlet 之间的跳转可以通过以下两种方式实现：

● 转向

转向（Forward）是通过 RequestDispatcher 类实现的，RequestDispatcher 可通过 request 的 getRequestDispatcher() 方法获得，getRequestDispatcher() 方法接收要跳转的路径，该路径必须以"/"开始，"/"表示 Web 程序的根目录，比如要跳转的 Servlet 为 http://localhost:8080/servlet/UserServlet.php，则参数应该为"/UserServlet.php"。Forward 允许跳转的路径包括 JSP 页面、Servlet，甚至是 WEB-INF 目录下的文件。下面是从 Servlet 跳转到另一个 Servlet 的代码。

```
RequestDispatcher requestDispatcher = request.getRequestDispatcher("/UserServlet");
requestDispatcher.forward(req, resp);
```

当使用 Forward 方式进行跳转时，地址栏会显示跳转前的访问地址，因为跳转过程是在服务器内部完成的，客户端并不知道，因此 Forward 跳转对于客户端浏览器是透明的。

- 重定向

重定向（Redirect）是通过 response 的 sendRedirect() 方法实现的，sendRedirect() 方法接收要跳转的路径，如果该路径以"/"开始，则"/"表示 Web 站点的根目录，比如要跳转的 Servlet 为 http://localhost:8080/servlet/UserServlet.php，则路径中的"/"表示 http://localhost:8080/。下面是从 Servlet 重定向到另一个 Servlet 的代码。

response.sendRedirect(request.getContextPath() + "/UserServlet");

当使用 Redirect 方式进行跳转时，地址栏会显示跳转后的访问地址，因为代码执行到 Redirect 时，会先回到客户端浏览器，再利用新地址发送新的请求，因此跳转是在客户端进行的。

⑥ MVC 框架

MVC 是 Model（模型）——View（视图）——Controller（控制器）的缩写，是一种软件设计规范，按照业务逻辑、数据、界面显示分离的方法来组织代码。三部分的分工如下：

Model 层，管理模块中用到的数据，负责实体类的定义和数据的存取，为应用程序提供数据支持。

View 层，提供模型的展示方式，是应用程序的外观。

Controller 层，接收用户提交的数据，调用 Model 层进行处理，并将结果推送给 View 层显示，是连接模型和视图的枢纽。

其中控制模块通过 Servlet 实现，视图模块通过 JSP 实现。

MVC 架构中各部分的关系如图 3.9 所示。

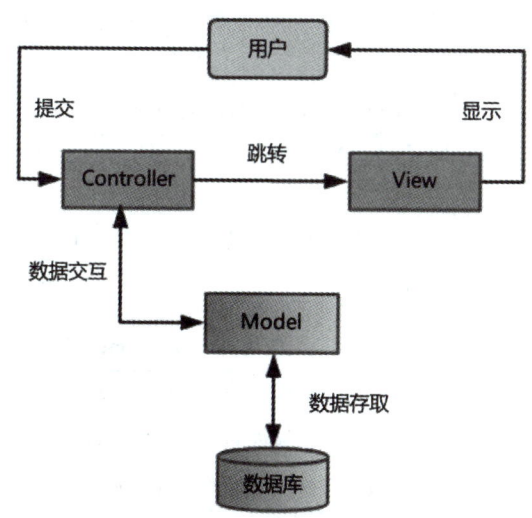

图 3.9　MVC 架构中各部分的关系

3.2.2　EL 表达式

EL（Expression Language）是一种表达式语言，其目的是为了使 JSP 书写起来更加简单。EL 表达式的语法结构为 ${expression}，EL 表达式是写在 JSP 的 HTML 代码中的，而不能写在包含"<%"与"%>"的 JSP 脚本中。在花括号中可以使用"[]"和"."两个运算符，当要存取的属性名称中包含一些特殊字符，如"."或"-"等并非字母或数字的符号，就一定要使用"[]"。例如 ${user.My-Name} 应当改为 ${user["My-Name"]}。如果要动态取值也必须使用"[]"来完成，而"."无法做到动态取值。例如 ${requestScope.user [data]} 中 data 是一个变量，此时必用"[]"来完成。

① EL 表达式

EL 表达式提供了获取对象及其属性的简单方法，例如 ${username} 表示取出某一范围中名称为 username 的变量。因为我们并没有指定哪一个范围的 username，所以它会依序从 page、request、session、application 范围中查找。假如途中找到 username，就直接返回该值，不再继续找下去，但如果全部的范围都没有找到就返回 null。EL 表达式也支持获取对象的属性，例如 ${user.name} 可以输出 user 对象的 name 属性，相当于调用 user.getName()。

② 隐藏对象

EL 表达式不仅可以读取 request、session 中的属性，还可以读取其他 JSP 隐藏对象的属性，EL 表达式中获取隐藏对象的方式详见表 3.9。

表 3.9　EL 表达式获取隐藏对象的方式

类别	标识符	描述
请求参数	param	获取参数值，返回类型为 String，如 ${param .foo} 等价于 request.getParameter("foo")
	paramValues	获取参数值数组，返回类型为 String[]。如请求地址为 user.jsp?name=tom&name=jane，此时使用 param 只能获取第一个值，而 paramValues 可以获取其他值，${paramValues.name[0]} 为 tom，${paramValues.name[1]} 为 jane
头信息	header	获取请求头信息，返回类型为 String，如 ${header.host} 可能返回 localhost:8080
	headerValues	获取头信息数组，返回类型为 String[]，如 ${headerValues. host [0]}
Cookie	cookie	获取 cookie 信息，如 cookie 为 ("username", "tom")，${cookie.username} 返回该 cookie 对象，${cookie. username.name} 返回"username"，${cookie. username.value} 返回"tom"
初始化参数	initParam	获取初始化参数，如 ${initParam.encoding}
作用域	pageScope	获取 page 作用域内的变量，如 ${pageScope.user.name}
	requestScope	获取 request 作用域内的变量，如 ${requestScope.user.name}
	sessionScope	获取 session 作用域内的变量，如 ${sessionScope.user.name}
	applicationScope	获取 application 作用域内的变量，如 ${applicationScope.user.name}

下面将在 JSP 中演示如何使用 EL 表达式获取变量的值，代码如下：

el.jsp

```jsp
<%@page import="java.util.*"%>
<%@page import="com.entity.*"%>
<%@page language="java" contentType="text/html; charset=UTF-8" pageEncoding="UTF-8"%>
<!DOCTYPE html PUBLIC "-//W3C//DTD HTML 4.01 Transitional//EN" "http://www.w3.org/TR/html4/loose.dtd">
<html>
<head>
<meta http-equiv="Content-Type" content="text/html; charset=UTF-8">
<title>el 表达式学习 </title>
</head>
<body>
   第一种，简单取值：
   <%
      String data = "hi,why!";
      request.setAttribute("data", data);
   %>
   ${data}
   <br>
   第二种，从 bean 中取值：
   <%
      User user = new User();
      user.setName("why");
      session.setAttribute("user", user);
   %>
   ${user.name}
   <br>
   第三种，从复杂 bean 中取值：
   <%
      User user1 = new User();
      Address address = new Address();
      address.setCity(" 上海 ");
      user1.setAddress(address);
      application.setAttribute("user1", user1);
   %>
   ${user1.address.city}
   <br>
   第四种，从 list 中取值：
   <%
      List userList = new ArrayList();
      userList.add(new User("session: why1"));
      userList.add(new User("session: why2"));
      session.setAttribute("userList", userList);
   %>
   ${userList[0].name }
   <br>
   第五种，从 map 中取值：
   <%
      Map map = new HashMap();
      map.put("why1", new User("request: why1"));
      map.put("why2", new User("request: why2"));
```

```
        request.setAttribute("map", map);
    %>
    ${map.why1.name}
    ${map['why2'].name}
    <br>
    获取当前应用的名称：
    ${pageContext.request.contextPath}
</body>
</html>
```

启动服务器测试上述代码，得到的结果如图 3.10 所示。

```
第一种,简单取值: hi,why!
第二种,从bean中取值: why
第三种,从复杂bean中取值: 上海
第四种,从list中取值: session: why1
第五种,从map中取值: request: why1 request: why2
获取当前应用的名称: /FirstProject
```

图 3.10 EL 表达式测试结果

③ EL 表达式运算

EL 表达式支持简单的运算、比较等操作，EL 表达式支持的运算如表 3.10 所示。

表 3.10 EL 表达式支持的运算类型

类型	描述
算术型	包括 +、-（二元）、*、/、div、%、mod、-（一元） 如 ${5 mod 2} 输出 1
逻辑型	包括 and、&&、or、||、!、not 如 ${not (1==2)} 输出 true
关系型	包括 ==、eq、!=、ne、<、lt、>、gt、<=、le、>=、ge 如 ${1 lt 2} 输出 true
条件型	如 ${1==2 ? 3 : 4} 输出 4
空值判断	对于数组或 Map 可以用 empty 判断是否为空 如 ${empty param.paramA} 判断 paramA 参数值是否为空或者个数为 0

下面将在 JSP 中演示 EL 表达式支持的操作，代码如下：

```
el.jsp
<%@page import="java.util.*"%>
<%@page language="java" contentType="text/html; charset=UTF-8" pageEncoding="UTF-8"%>
<!DOCTYPE html PUBLIC "-//W3C//DTD HTML 4.01 Transitional//EN" "http://www.w3.org/TR/html4/loose.dtd">
<html>
<head>
<meta http-equiv="Content-Type" content="text/html; charset=UTF-8">
<title>el 表达式学习 </title>
</head>
<body>
```

```
第一种 , 算术型 :<br>
5 + 2 = ${5 + 2},
5 * 2 = ${5 * 2},
5 div 2 = ${5 div 2},
5 mod 2 = ${5 mod 2}
<br>
第二种 , 逻辑型 :<br>
(not (1==2)) = ${not (1==2)},
((1==2) && (true)) = ${(1==2) && (true)}
<br>
第三种 , 关系型 :<br>
(1 lt 2) = ${1 lt 2},
(1 == 2) = ${1 lt 2}
<br>
第四种 , 条件型 :<br>
(1==2 ? 3 : 4) = ${1==2 ? 3 : 4}
<br>
第五种 , 判断 map 是否为空 :<br>
<%
    Map map = new HashMap();
%>
map : ${empty map}
<br>
</body>
</html>
```

启动服务器测试上述代码，得到的结果如图 3.11 所示。

```
第一种,算术型:
5 + 2 = 7, 5 * 2 = 10, 5 div 2 = 2.5, 5 mod 2 = 1
第二种,逻辑型:
(not (1==2)) = true, ((1==2) && (true)) = false
第三种,关系型:
(1 lt 2) = true, (1 == 2) = true
第四种,条件型:
(1==2 ? 3 : 4) = 4
第五种,判断map是否为空:
map : true
```

图 3.11 EL 表达式支持的操作测试结果

【任务实施】

1. 新建项目并配置 web.xml 文件

选择菜单"File"—"New"—"Dynamic Web Project"（如果没有列出，在"Other…"中查找），在弹出的窗口中填写"Project name"为"servlet"，"Dynamic Web module version"选择"2.5"，点击"Finish"按钮创建新项目。

在 web.xml 中修改 Web 应用程序的主页设置，将主页改为 reg.jsp，并配置 Servlet，具体代码如下：

```xml
web.xml
<welcome-file-list>
  <welcome-file>reg.jsp</welcome-file>
</welcome-file-list>
<servlet>
  <servlet-name>UserServlet</servlet-name>
  <servlet-class>com.servlet.UserServlet</servlet-class>
  <init-param>
    <param-name>encoding</param-name>
    <param-value>utf-8</param-value>
  </init-param>
  <load-on-startup>1</load-on-startup>
</servlet>
<servlet-mapping>
  <servlet-name>UserServlet</servlet-name>
  <url-pattern>/UserServlet</url-pattern>
</servlet-mapping>
```

2. 编写注册页面

新建 reg.jsp 页面，并输入如下代码：

```jsp
reg.jsp
<%@ page language="java" contentType="text/html; charset=UTF-8" pageEncoding="UTF-8"%>
  <!DOCTYPE html PUBLIC "-//W3C//DTD HTML 4.01 Transitional//EN" "http://www.w3.org/TR/html4/loose.dtd">
  <html>
<head>
  <meta http-equiv="Content-Type" content="text/html; charset=UTF-8">
  <title> 用户注册 </title>
</head>
<body>
<form action="UserServlet" method="post">
  <h2> 用户注册 </h2><br>
  用户名 <input type="text" name="uid"><br>
  密　码 <input type="password" name="pwd"><br>
  性　别 <select name="sex">
    <option value="M"> 男 </option>
    <option value="F"> 女 </option>
    </select><br>
  生　日 <input type="text" name="birthday"><br>
  爱　好
  <input type="checkbox" name="interest" value="1"> 运动
  <input type="checkbox" name="interest" value="2"> 音乐
  <input type="checkbox" name="interest" value="3"> 旅游 <br><br>
  <input type="submit" value=" 注册 ">
</form>
</body>
</html>
```

3. 编写 UserServlet 类

UserServlet 中会获取注册页面提交的参数，并在 UserServlet 中使用 3.1.3 节中的实体类、Service 层和 DAO 层代码，代码如下：

UserServlet.java

```java
package com.servlet;
public class UserServlet extends HttpServlet {
  @Override
   protected void doGet(HttpServletRequest request, HttpServletResponse response) throws ServletException, IOException {
      this.doPost(request, response);
   }
  @Override
   protected void doPost(HttpServletRequest request, HttpServletResponse response) throws ServletException, IOException {
      String encoding = getInitParameter("encoding"); // 获取 web.xml 中配置的参数
      request.setCharacterEncoding(encoding);    // 为 request 指定编码格式
      response.setCharacterEncoding(encoding);    // 为 response 指定编码格式
      String name = request.getParameter("name"); // 获取页面提交的数据
      String password = request.getParameter("password");
      String sex = request.getParameter("sex");
      String birthday = request.getParameter("birthday");
      String[] interests = request.getParameterValues("interest");
      String endInterests = "";
      String error = "";
      if(null != interests) {
        for (String interest : interests) {
           endInterests += interest + ",";
        }
        endInterests = endInterests.substring(0, endInterests.length()-1);
      }

      Pattern p = Pattern.compile("^[A-Za-z0-9]{3,6}$");
      if(!p.matcher(name).find()) { // 验证用户名格式
        error = " 用户名格式不合法（要求由 3-6 个字母和数字组成）";
        request.setAttribute("error", error);
        request.getRequestDispatcher("/result.jsp").forward(request, response);
        return;
      }
      p=Pattern.compile("^[A-Za-z0-9]{6,10}$");
      if(!p.matcher(password).find()) { // 验证密码格式
        error = " 密码格式不合法（要求由 6-10 个字母和数字组成）";
        request.setAttribute("error", error);
        request.getRequestDispatcher("/result.jsp").forward(request, response);
        return;
      }
      if(!isDate(birthday)) { // 验证生日格式
        error = " 生日格式不合法 ";
        request.setAttribute("error", error);
        request.getRequestDispatcher("/result.jsp").forward(request, response);
```

```java
        return;
    }
    
    User user = new User(name, password, sex, birthday, endInterests);
    UserService service = new UserService();
    int result = service.register(user);  // 执行注册操作
    if(-1 == result) {  // 用户名已经存在
        error = " 用户名已经存在 ";
        request.setAttribute("error", error);
        request.getRequestDispatcher("/result.jsp").forward(request, response);
        return;
    } else if(0 == result){  // 注册失败
        error = " 注册失败 ";
        request.setAttribute("error", error);
        request.getRequestDispatcher("/result.jsp").forward(request, response);
        return;
    } else {  // 注册成功
        request.setAttribute("user", user);
        request.getRequestDispatcher("/result.jsp").forward(request, response);
        return;
    }
}

boolean isDate(String s) {
    SimpleDateFormat df=new SimpleDateFormat("yyyy-MM-dd");
    df.setLenient(false); // 设置为严格验证
    try{
        df.parse(s);
        return true;
    }catch (Exception e){
        return false;
    }
}
}
```

4. 编写注册信息显示页面

新建 result.jsp，并在页面中使用 EL 表达式显示用户信息，代码如下：

result.jsp

```jsp
<%@ page language="java" contentType="text/html; charset=UTF-8" pageEncoding="UTF-8"%>
<!DOCTYPE html PUBLIC "-//W3C//DTD HTML 4.01 Transitional//EN" "http://www.w3.org/TR/html4/loose.dtd">
<html>
<head>
<meta http-equiv="Content-Type" content="text/html; charset=UTF-8">
<link rel="stylesheet" type="text/css" href="css/style.css">
<title> 结果页面 </title>
</head>
<body>
    提交的用户名是：${user.name}<br>
```

  提交的密码是：${user.password}
  提交的性别是：${user.sex}
  提交的生日是：${user.birthday}
  提交的爱好是：${user.interest}
  <div class = "error">${error}</div>
</body>
</html>

5. 运行项目

  点击工具栏上的"Run"按钮（快捷键 Ctrl+F11），在弹出的窗口中选择"Run On Server"，启动完成后在浏览器中输入"http://localhost:8080/servlet"打开页面，效果如图 3.12 所示。

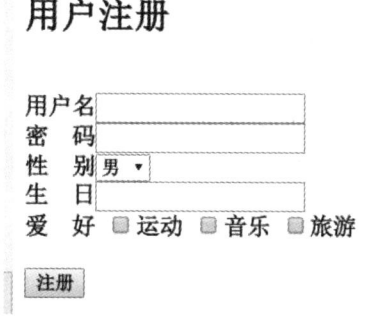

图 3.12 用户注册页面效果

  如果用户填写的注册信息没有验证通过，程序不会向数据库中插入记录，同时会为用户响应失败信息，注册失败页面效果如图 3.13 所示。

  提交的用户名是：
  提交的密码是：
  提交的性别是：
  提交的生日是：
  提交的爱好是：
  用户名格式不合法（要求由3-6个字母和数字组成）

图 3.13 注册失败页面效果

  在显示用户注册信息的时候，用到了 EL 表达式，注册成功页面效果如图 3.14 所示。

  提交的用户名是：tom
  提交的密码是：123456
  提交的性别是：F
  提交的生日是：1991-09-09
  提交的爱好是：1,2,3

图 3.14 注册成功页面效果

【实践训练】

完善用户信息，增加电话和邮箱属性，在 Servlet 中对用户信息进行合法性验证，并将信息添加到数据库中，然后将用户注册信息展示给用户。

拓展训练

查阅资料，实现修改用户注册信息功能，具体要求如下：

1．功能
（1）用户登录，查看原有注册信息。
（2）填写新的信息并提交。
（3）验证新的注册信息，通过验证则保存，否则给出错误提示。

2．实现方式
（1）用 Servlet 结合 JSP 中的 EL 表达式显示用户注册信息。
（2）在 Servlet 中用正则式对提交的新注册信息进行合法性验证。
（3）采用分层模式实现数据库的操作。

同步训练

一、填空题

1．Java 操作 JDBC 的接口位于 _____ 包内。
2．JDBC 中 _____ 类用于访问数据库查询结果。
3．HTTP 提供了多种访问方式，分别为 _____、_____、HEAD、DELETE、TRACE、PUT 和 OPTIONS。
4．Servlet 之间的跳转有 _____ 和 _____ 两种方式。
5．EL 表达式的语法结构为 _____。

二、简答题

1．Statement 和 PreparedStatement 的主要区别是什么？
2．DAO 模式的优势是什么？
3．简述编写 Servlet 的步骤。
4．简述 Servlet 的生命周期。
5．简述 MVC 模式中的 M、V 和 C 分别代表什么，MVC 模式有什么优势。

项目 4
实现用户管理功能

单元介绍

　　通过用户管理项目，掌握 Cookie 和 Session 这两种会话跟踪机制的使用，以及用 EL 表达式和 JSTL 标签实现 JSP 中数据显示的方法。
　　本项目包括以下任务：
- 分别使用 Cookie 和 Session 实现会话跟踪
- 掌握 JSTL 常用标签的用法
- JSTL 标签结合 EL 表达式在 JSP 中实现数据显示

学习目标

【知识目标】
　　掌握 HttpSession 的常用方法
　　掌握 Cookie 的常用属性
　　掌握 JSTL core 标签库中的常用标签
【能力目标】
　　能正确使用 Session 实现会话跟踪
　　能正确使用 Cookie 实现会话跟踪
　　能正确使用 EL 表达式和 JSTL 标签在 JSP 中实现数据显示

任务 4.1　管理员登录

【任务分析】

管理员登录功能的执行过程是，首先要进入登录界面，输入并提交数据，服务器端程序对提交的用户名和密码进行合法性验证，若不通过，需要提示用户，若合法性验证通过，则需通过查询数据库对用户输入的用户名和密码进行验证，验证失败需要提示用户，验证成功则将用户信息存入 Session 中，并显示登录成功界面。

【相关知识】

4.1.1　Session 和 Cookie

Web 应用程序使用 HTTP 传输数据，而 HTTP 是无状态的协议，一旦数据传输完毕，客户端与服务器的连接就会关闭，如果客户端需要再次发出请求，则需要建立新的连接，这就意味着无法达到跟踪会话的目的。当用户登录验证完毕后，服务器和客户端连接断开，再继续访问其他页面时，服务器不能"记住"该用户刚才曾登录过。为解决服务器"失忆"的问题，需要使用技术手段对用户连接上服务器后的所有会话进行跟踪，以弥补 HTTP 无状态的不足。常用的会话跟踪技术是借助 Session 和 Cookie，Session 在服务器端保存用户信息，相当于一个储物柜，然后将储物柜的钥匙（Session ID）发放给客户端，并存放在 Cookie 中，用户每次发送请求，都将 Session ID 提交上来，以确认身份。

① Cookie 工作机制

Cookie 原意为"甜饼"或"饼干"，由 W3C 组织提出，最早是由 Netscape 社区发展的一种客户端数据保存机制。目前所有的主流浏览器都支持 Cookie。

Cookie 实际上是在文件中存储键值对。客户端向服务器发送请求，若服务器需要记录用户的状态，就使用 Response 对象向客户端发送一个 Cookie 对象，客户端浏览器就会把该 Cookie 对象保存起来，当再次请求服务器时，会把 Cookie 的内容连同要提交的信息一并发给服务器，服务器就可以根据 Cookie 信息来辨别用户身份了。

想要查看某个网站的 Cookie，只需要在浏览器地址栏中输入"javascript:alert (document.cookie)"，再按回车键就可以了，图 4.1 展示了百度页面的 Cookie 信息。

● Cookie 类

Cookie 被封装到 javax.servlet.http.Cookie 类中，在 JSP 中不需要引入 Cookie 类就可以直接使用，服务器端程序通过 request.getCookie() 获取客户端提交的所有 Cookie，通过 response.addCookie(Cookie cookie) 向客户端设置 Cookie。

图 4.1　百度页面的 Cookie 信息

Cookie 支持多种属性，并为每种属性提供了 getter 和 setter 方法，Cookie 中提供的常用属性如表 4.1 所示。

表 4.1　Cookie 的常用属性

属性名：类型	属性描述
name：String	Cookie 的名称，Cookie 一旦创建，名称便不能修改
value：Object	Cookie 的值，若值为 Unicode 字符，需要为字符编码，如果值为二进制数据，则需要使用 BASE64 编码
maxAge：int	Cookie 的失效时间，单位为秒，如果为正数，则在 maxAge 秒之后失效，如果为 0 表示删除该 Cookie，如果为负数，该 Cookie 为临时 Cookie，关闭浏览器后便失效。默认为 -1
secure：boolean	Cookie 是否使用安全协议传输，若为 true 表示需要使用安全协议，否则不使用，默认为 false，安全协议有 HTTPS、SSL 等
path：String	Cookie 的使用路径，如果设置为 "/servlet/"，则只有 contextPath 为 "/servlet" 的程序可以访问该 Cookie，如果设置为 "/"，则本域名下的 contextPath 均可访问该 Cookie，注意最后一个字符必须为 "/"
domain：String	可以访问该 Cookie 的域名，如果设置为 ".baidu.com"，则所有以 ".baidu.com" 结尾的域名均可访问该 Cookie，注意第一个字符必须为 "."
comment：String	Cookie 的说明，浏览器显示 Cookie 信息的时候显示该说明
version：int	该 Cookie 的版本号，0 表示遵循 Netscape 的 Cookie 规范，1 表示遵循 W3C 的 RFC2109 规范

下面将通过一个程序来演示各个属性的用途，包括如何设置 Cookie 值，如何改变 Cookie 值，以及如何删除 Cookie。

cookie.jsp
```jsp
<%@ page language="Java" import="java.util.*" pageEncoding="UTF-8" import="java.net.URLEncoder"%>
<!DOCTYPE HTML PUBLIC "-//W3C//DTD HTML 4.01 Transitional//EN">
<%!
boolean isNull(String str){  // 判断字符串是否为空
   return str == null||str.trim().length()==0;
   }
%>
<%
   request.setCharacterEncoding("UTF-8");          // 设置 request 编码
   response.setCharacterEncoding("UTF-8");         // 设置 response 编码
   String name = request.getParameter("name");     // 获取各个属性的值
   String value = request.getParameter("value");
   String maxAge = request.getParameter("maxAge");
   String domain = request.getParameter("domain");
   String path = request.getParameter("path");
   String comment = request.getParameter("comment");
   String secure = request.getParameter("secure");
if(!isNull(name))
{
// 创建 Cookie
      Cookie cookie = newCookie(URLEncoder.encode(name,"UTF-8"),URLEncoder.encode(value,"UTF-8"));
 if(!isNull(maxAge)) // 设置 Cookie 的有效时间
   cookie.setMaxAge(Integer.parseInt(maxAge));
 if(!isNull(domain)) // 设置可访问 Cookie 的域名
   cookie.setDomain(domain);
 if(!isNull(path)) // 设置可访问 Cookie 的路径
   cookie.setPath(path);
 if(!isNull(comment)) // 设置 Cookie 的说明
   cookie.setComment(comment);
 if(!isNull(secure)) // 设置是否使用安全协议传输数据
   cookie.setSecure("true".equalsIgnoreCase(secure));
 response.addCookie(cookie);
}
%>
<html>
<head>
 <title>Cookie 属性实例 </title>
 <meta http-equiv="pragma" content="no-cache">
 <meta http-equiv="cache-control" content="no-cache">
 <meta http-equiv="expires" content="0">
</head>
<body>
<div>
 <fieldset>
  <legend> 当前有效的 Cookie</legend>
   <script type="text/JavaScript">
    document.write(document.cookie);
```

```
    </script>
   </fieldset>
   <fieldset>
    <legend>设置新的 Cookie</legend>
     <form action="cookie.jsp" method="post">
      <table>
       <tr>
        <td>name:</td>
        <td><input type="text" name="name" style="width:200px;"/></td>
       </tr>
       <tr>
        <td>value:</td>
        <td><input type="text" name="value" style="width:200px;"/></td>
       </tr>
       <tr>
        <td>maxAge:</td>
        <td><input type="text"name="maxAge"style="width:200px;"/></td>
       </tr>
       <tr>
        <td>domain:</td>
        <td><input type="text" name="domain"style="width:200px;"/></td>
       </tr>
       <tr>
        <td>path:</td>
        <td><input type="text" name="path" style="width:200px;"/></td>
       </tr>
       <tr>
        <td>comment:</td>
        <td><input type="text" name="comment" style="width:200px;"/></td>
       </tr>
       <tr>
        <td>secure:</td>
        <td><input type="text" name="secure" style="width:200px;"/></td>
       </tr>
       <tr><td></td>
<td>
<input type="submit" value=" 提交 "/>
        <input type="button" value=" 刷新 " onclick="window.location='setCookie.jsp'"/>
</td>
</tr>
</table>
</form>
</fieldset>
</div>
</body>
</html>
```

程序的运行结果如图 4.2 所示。

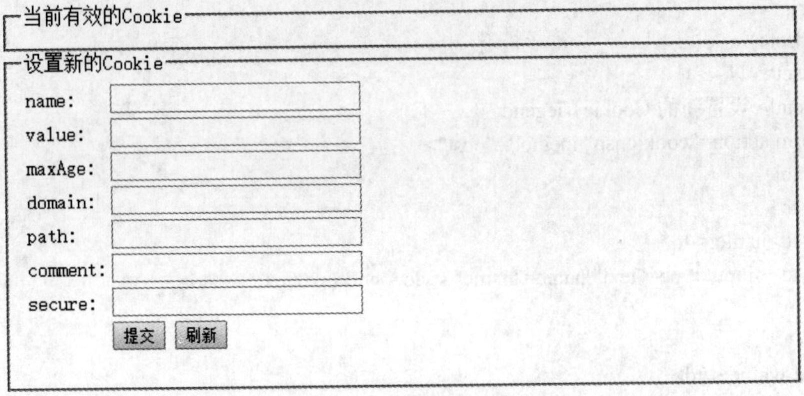

图 4.2　设置 Cookie 属性值运行结果（1）

当在 name 和 value 文本框内输入值并提交之后，运行结果如图 4.3 所示。

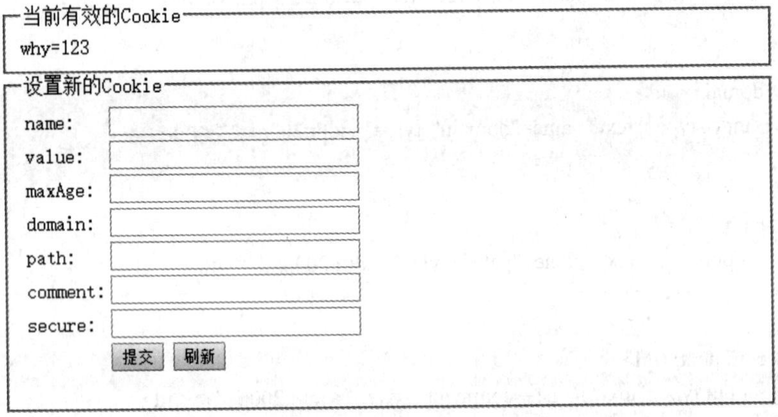

图 4.3　设置 Cookie 属性值运行结果（2）

● Cookie 的域名

很多网站都会使用 Cookie，不同网站之间的 Cookie 是不能共享的，比如百度网站只能访问百度的 Cookie，而不能够访问和修改谷歌的 Cookie，这是因为 Cookie 具有不可跨域性。Cookie 在客户端是由浏览器进行管理的，而浏览器判断一个网站能否操作 Cookie 的依据是域名。

正因为 Cookie 的不可跨域性，增强了 Cookie 的安全性，但同一个一级域名下面的两个二级域名虽然域名不同，但有时需要共享 Cookie，此时可以通过设置 domain 属性来达到共享的目的，代码如下所示。

```
Cookie cookie = new Cookie("name", "why");     // 新建 Cookie
cookie.setDomain(".hello.com");                // 设置域名
```

domain 参数必须以 "." 开始，name 属性相同，但是 domain 属性不同的两个 Cookie 是不同的 Cookie。

● Cookie 的路径

domain 属性决定着可以访问 Cookie 的域名，而 path 属性则决定着允许访问 Cookie 的路径（ContextPath），例如下面的程序实现了只允许 /servlet/ 下的路径可以访问该 Cookie，具体的代码如下所示。

```
Cookie cookie = new Cookie("name", "why");    // 新建 Cookie
cookie.setPath("/servlet/");                  // 设置路径
```

若设置路径为"/"，则允许所有路径均可访问该 Cookie，path 属性必须以"/"结尾。两个 name 相同，但 path 不同的 Cookie 也是不同的，页面只能获取它所属 path 的 Cookie。

● Cookie 的安全属性

HTTP 不仅是无状态的，而且也是不安全的，在不安全的协议中传输 Cookie，容易被截获，因此，若不希望 Cookie 在不安全的协议中传输，可以设定 secure 属性为 true，此时 Cookie 只能在安全协议中传输，比如 HTTPS、SSL 等。具体的代码如下所示。

```
Cookie cookie = new Cookie("name", "why");    // 新建 Cookie
cookie.setSecure(true);                       // 设置安全属性
```

● Cookie 中保存中文

中文字符和英文字符不同，中文属于 Unicode 字符，在内存中占 4 字节，而英文字符属于 ASCII 字符，在内存中占 2 字节，在 Cookie 中使用中文字符时，需要对其进行编码，否则会产生乱码。编码可以使用 java.net.URLEncoder 类中的 encode(String str, String encoding) 方法，解码可以使用 java.net.URLDecoder 类的 decode(String str, String encoding) 方法。

下面是使用 UTF-8 对中文进行编码的代码：

```
Cookie cookie = newCookie(URLEncoder.encode("姓名","UTF-8"),
        URLEncoder.encode("张三","UTF-8"));
```

下面是使用 UTF-8 对中文进行解码的代码：

```
String name = URLDecoder.decode(cookie.getName(), "UTF-8");
String value = URLDecoder.decode(cookie.getValue(), "UTF-8");
```

在对中文进行编码时，一般使用 UTF-8 编码格式，不推荐使用 GBK 编码方式，浏览器不一定支持，而且 JavaScript 也不支持 GBK 编码。

● Cookie 的有效期

Cookie 的有效期是通过 maxAge 属性来决定的，其单位为秒。

若 maxAge 的值为正数，则表示 Cookie 在指定时间之后失效，浏览器会将其持久化到文件中，即使关闭浏览器之后，只要在 maxAge 时间之内，Cookie 仍然有效。下面的代码将使 Cookie 永久有效。

```
Cookie cookie = new Cookie("name", "why"); // 新建 Cookie
Cookie.setMaxAge(Integer.MAX_VALUE);
```

若 maxAge 的值为负数，表示该 Cookie 为临时 Cookie，存在于浏览器内存中，只有在本浏览器窗口及其子窗口内有效，若关闭浏览器之后，该 Cookie 就会失效，Cookie 的 maxAge 的值默认为 -1。

若 maxAge 为 0，表示删除该 Cookie。Cookie 没有提供删除方法，因此可通过设置 Cookie 的 maxAge 为 0 的方法实现删除 Cookie 的目的。

从客户端读取 Cookie 时，浏览器只会提交 name 和 value 属性，其他属性是不可读的，maxAge 属性仅供浏览器判断 Cookie 是否失效。

- Cookie 的修改和删除

Cookie 并没有提供修改和删除方法，要先修改某个 Cookie，只需要建立一个同名的 Cookie，并添加到 response 中覆盖原来的 Cookie 即可。要想删除某个 Cookie，只需要建立一个同名的 Cookie，并将 maxAge 属性设置为 0，并添加到 response 中覆盖原来的 Cookie 即可。

修改和删除 Cookie 时，新建的 Cookie 除了 value 和 maxAge 属性外，其他诸如 name、domain 和 path 等属性必须与原 Cookie 完全相同，否则浏览器会将认为是两个不同的 Cookie，导致修改或删除失败。

② Session 工作机制

Session 是服务器端为每个用户单独开辟的的一个私有存储区，默认情况下是将数据存放在服务器内存中，也可以设置为存放到文件或数据库中。每个 Session 都有独一无二的 SessionID，服务器端程序通过 SessionID 来获取 Session 中存放的数据。

- 获取 session 对象

通过 request 对象的 getSession() 方法，可以获取到 session 对象，然后用 session 对象的方法进行具体操作。

- session 的常用方法

session 对象所属类实现了 javax.servlet.http.HttpSession 接口，接口中规定的常用操作方法如表 4.2 所示。

表 4.2　HttpSession 的常用方法

方法名	方法描述
Object getAttribute(String name)	返回名称为 name 的属性值
long getCreationTime()	返回 Session 的创建日期
String getId()	返回 Session 的 ID，该 ID 由服务器自动创建
long getLastAccessedTime()	返回 Session 的最后活跃时间
int getMaxInactiveInterval()	返回 Session 的超时时间，单位为秒
void setMaxInactiveInterval(int interval)	设置 Session 的超时时间，单位为秒
void invalidate()	使 Session 失效
void removeAttribute(String name)	移除名称为 name 的属性
void setAttribute(String name, Object value)	设置 Session 属性

用户登录系统成功之后，通常将用户信息存放到 Session 中，核心代码如下：

HttpSession session = request.getSession(); // 获取 Session 对象
session.setAttribute("user", user);　　　　// 将 user 放到 Session 中

将用户信息放到 Session 中之后，就可以在其他 JSP 页面或 Servlet 中获取当前登录用户的信息，核心代码如下：

```
HttpSession session = request.getSession();        // 获取 Session 对象
User user = (User)session.getAttribute("user");    // 获取 Session 中的 user 对象
```

● Session 的生命周期

Session 是保存在服务器端的，为了获取更高的存取速度，服务器一般把 Session 放到内存中，每个用户都会有一个独立的 Session。如果 Session 过于复杂，会导致服务器内存溢出，因此应保证 Session 中的内容尽量精简。

Session 在用户第一次访问服务器时创建，只有访问 JSP、Servlet 等程序时，才会创建 Session，访问 HTML 等静态资源时，是不会创建 Session 的，如果尚未生成 Session，可以使用 request.getSession(true) 强制生成 Session。

在生成 Session 之后，只要用户访问了服务器，不管是否使用 Session，服务器都会更新 Session 的最后访问时间。

由于 Session 存在于服务器内存中，当有大量用户访问时，随着 Session 数量的增多，服务器容易产生内存溢出，为了尽量避免这类问题，服务器会把长时间没有被访问的 Session 从内存中删除，这个时间就是 Session 的超时时间。可调用 Session 的 setMaxInactiveInterval 方法设置超时时间，也可在 web.xml 中配置 Session 的超时时间，具体代码如下：

```
<session-config>
<session-timeout>15</session-timeout><!-- 单位为分钟 -->
</session-config>
```

对于不再使用的 Session，可以调用 Session 的 invalidate 方法使 Session 立即失效。

● URL 地址重写

虽然 Session 是保存在服务器的，但是 Session 的正常运行仍然需要客户端的支持，这是因为 Session 不能通过 HTTP 区分用户，它是根据 Cookie 来标识用户的，Session 会向客户端发送一个名为 JSESSIONID 的 Cookie，它的值为该 Session 的 ID，Session 会依据该 Cookie 来标识是否为同一个用户。

该 Cookie 为服务器自动生成的，它的 maxAge 属性值为 -1，这就意味着该 Cookie 是存在于浏览器内存中，不会被持久化到文件中，当浏览器关闭之后，该 Cookie 就会被清除。同一个客户端的两个浏览器访问服务器时，会生成两个不同的 Session，但是在浏览器内部窗口打开链接，子窗口会共享父窗口的 Cookie，因此也会共享一个 Session。

如果客户端浏览器禁用了 Cookie 功能，或者根本不支持 Cookie，此时若还想使用 Session 功能，就只能重写 URL 地址了。URL 地址重写的原理是将客户端 Session 的 ID 重写到 URL 地址中，服务器接收到这样的地址后，就可解析出 Session 的 ID，这样即使客户端不支持 Cookie，服务器也可以区分用户身份。HttpServletResponse 类提供了 encodeURL(String url) 实现 URL 地址重写，核心代码如下：

```
response.encodeURL("index.jsp?name=why&pwd=123");
```

encodeURL 方法会判断客户端是否支持 Cookie，若支持,则将地址原封不动的返回，否则会将 Session 的 ID 写到 URL 中，重写后的 URL 如下所示。

```
index.jsp;jsessionid=5F4771183629C9834F8382E23BE13C4C?name=why&pwd=123
```

可以观察到，在文件名的后面，请求参数的前面被加上了";jsessionid=XXX"，其中XXX为Session的ID，添加的部分既不会影响请求的文件资源，也不会影响请求参数。服务器在接收到地址后，就可以解析出Session的ID，从而获取到存放在Session中的数据。

如果是页面重定向，可以使用response的encodeRedirectURL(String url)方法实现地址重写，核心代码如下：

response.sendRedirect(response.encodeRedirectURL ("index.jsp"));

当用户浏览器不支持Cookie或者浏览器的Cookie功能被禁用时，可以禁止Session使用Cookie，统一使用URL地址重写。可以通过配置的方式实现Session禁用Cookie，具体有如下两种方式。

第一种方式，在WebContent目录下的META-INF文件夹内打开context.xml文件（若没有，则新建），内容如下：

```
<?xml version="1.0" encoding="UTF-8"?>
<Context path= "/servlet/" cookies="false">
</Context>
```

第二种方式，修改Tomcat目录下的conf/context.xml，修改内容如下：

```
<Context cookies="false">
</Context>
```

这两种配置方式只是禁止Session使用Cookie作为识别标志，但是程序仍然可以使用其他的Cookie。

③ Cookie与Session的比较

Cookie和Session都是实现会话跟踪的技术，一般情况下，两者均能满足需求，但是在某些特定情况下，只能用Cookie或Session，下面将从不同的角度详细比较两者的适用场景。

● 从存取数据上比较

Cookie中只能存储ASCII字符，如果需要存储Unicode字符或者二进制数据，需要对数据进行UTF-8或BASE64等方式编码，Cookie中也不能直接存取Java对象。

Session中可以存取任何类型的数据，比如Java基本数据类型，Java集合类型，甚至是自定义的Java类型。

● 从隐私安全上比较

Cookie是存储在客户端浏览器中，Cookie信息可能会被客户端截获或修改，而Session是存储在服务器的，对客户端是透明的，因此可以存储任何隐私信息。想要解决Cookie信息容易泄露的问题，可以对敏感信息进行加密传输，服务器拿到信息后再进行解密，保证Cookie信息的安全。而使用Session的话，就免去了这些麻烦。

● 从有效期上比较

Cookie的有效期设置比较灵活，可以根据需要对maxAge属性值进行设置，若将其设置为Integer.Max_Value，则此Cookie为永久有效；若为正数，则在规定时间内都有效，而不受浏览器是否关闭的影响；若为0，则表示删除此Cookie；若为负数，则当浏览器关闭后，此Cookie失效。

Session 的有效期可以通过属性 MaxInactiveInterval 的值进行设置，但是 Session 依赖于名为 JSESSIONID 的 Cookie，而该 Cookie 的 maxAge 默认为 -1，意味着只要浏览器关闭了该 Session 就会失效，因此 Session 不能实现长时间免登录的效果。而且 Session 的有效期设置越长，服务器存在的 Session 数量就会越多，容易导致服务器内存溢出。

- 从对服务器负担上比较

Cookie 并不占用服务器资源，它是保存在客户端的，因此并发浏览的用户比较多时，Cookie 是比较好的选择。

Session 是存储在服务器的，占用服务器资源，因此当用户访问量非常大时，会消耗大量的内存，此时使用 Session 对用户进行会话跟踪不是好的选择。

- 从浏览器支持上比较

Cookie 需要浏览器的支持，并且需要浏览器开启 Cookie 功能，否则不能使用 Cookie 进行会话跟踪。此时只能使用 Session 和 URL 地址重写技术进行会话跟踪。

- 从跨域名上比较

Cookie 支持跨域名访问，例如将 domain 设置为 ".why.com"，则以 ".why.com" 为结尾的域名均可访问此 Cookie，而 Session 则不支持跨域名访问，Session 仅在它所在的域名内有效。

在某些情况下，仅使用 Session 或仅使用 Cookie 并不能满足需求，需要将两者搭配使用，才能达到最终的目的。

【任务实施】

1. 新建项目并配置 web.xml 文件

选择菜单"File"—"New"—"Dynamic Web Project"（如果没有列出，在"Other…"中查找），在弹出的窗口中填写"Project name"为"servlet"，连续点击"Next"两次，在"Web Module"选项卡中勾选"Generate web.xml deployment descriptor"复选框，让项目中自动添加 web.xml 文件，最后点击"Finish"按钮创建新项目。

在 web.xml 中修改 Web 应用程序的主页设置，将主页改为 login.jsp，并配置 Servlet，具体代码如下：

```
web.xml
<welcome-file-list>
  <welcome-file>login.jsp</welcome-file>
</welcome-file-list>
<servlet>
  <servlet-name>UserServlet</servlet-name>
  <servlet-class>com.servlet.UserServlet</servlet-class>
  <init-param>
    <param-name>encoding</param-name>
    <param-value>utf-8</param-value>
```

```xml
    </init-param>
    <load-on-startup>1</load-on-startup>
</servlet>
<servlet-mapping>
    <servlet-name>UserServlet</servlet-name>
    <url-pattern>/UserServlet</url-pattern>
</servlet-mapping>
```

2. 新建并编辑登录页面

新建 login.jsp 页面,并输入如下代码:

login.jsp
```jsp
<%@ page language="java" contentType="text/html; charset=UTF-8" pageEncoding="UTF-8"%>
<!DOCTYPE html PUBLIC "-//W3C//DTD HTML 4.01 Transitional//EN" "http://www.w3.org/TR/html4/loose.dtd">
<html>
<head>
    <meta http-equiv="Content-Type" content="text/html; charset=UTF-8">
    <link rel="stylesheet" type="text/css" href="css/style.css">
    <title> 用户登录 </title>
</head>
<body>
 <form action="UserServlet?flag=login" method="post">
    <h2> 用户登录 </h2><br>
    用户名 <input type="text" name="name"><br>
    密    码 <input type="password" name="password"><br>
    <div class = "error">${error}</div>
    <br><br>
    <input type="submit" value=" 登录 ">
 </form>
</body>
</html>
```

3. 编写 UserServlet 类

UserServlet 中会获取登录页面提交的参数,并在 UserServlet 中使用 3.1.3 节中的实体类,具体代码如下:

UserServlet.java
```java
package com.servlet;
public class UserServlet extends HttpServlet {
    private UserService service = new UserService();
    @Override
    protected void doGet(HttpServletRequest request, HttpServletResponse response) throws ServletException, IOException {
        this.doPost(request, response);
    }
    @Override
    protected void doPost(HttpServletRequest request, HttpServletResponse response) throws ServletException, IOException {
        String encoding = getInitParameter("encoding");         // 获取 web.xml 中的初始化参数
        request.setCharacterEncoding(encoding);                 // 设置 request 的编码格式
```

```java
            response.setCharacterEncoding(encoding);        // 设置 response 的编码格式
            String flag = request.getParameter("flag");     // 获取操作类型
            if(null != flag && flag.equals("reg")) {        // 注册
                reg(request, response);
                return;
            }
        else if(null != flag && flag.equals("login")) {     // 登录
                try {
                    login(request, response);
                }
        catch (SQLException e) {
                    e.printStackTrace();
                    return;
                }
                return;
            }
        }

        // 登录验证
        private void login(HttpServletRequest request, HttpServletResponse response) throws
ServletException, IOException, SQLException {
            String name = request.getParameter("name");
            String password = request.getParameter("password");
            String error = "";

            Pattern p = Pattern.compile("^[A-Za-z0-9]{3,6}$");
            if(!p.matcher(name).find()) {
                error = " 用户名格式不合法（要求由 3-6 个字母和数字组成）";
                request.setAttribute("error", error);
                request.getRequestDispatcher("/login.jsp").forward(request, response);
                return;
            }
            p=Pattern.compile("^[A-Za-z0-9]{6,10}$");
            if(!p.matcher(password).find()) {
                error = " 密码格式不合法（要求由 6-10 个字母和数字组成）";
                request.setAttribute("error", error);
                request.getRequestDispatcher("/login.jsp").forward(request, response);
                return;
            }

            User user = new User(name, password);
            User u = service.login(user);
            if(null == u) {
                error = " 用户名或者密码错误 ";
                request.setAttribute("error", error);
                request.getRequestDispatcher("/login.jsp").forward(request, response);
                return;
            } else {
                HttpSession session = request.getSession();          // 获取 Session 对象
                session.setAttribute("user", u);                     // 将 user 放到 Session 中
```

```java
            request.setAttribute("userList", service.queryUsers(null));// 设置列表给页面
            request.getRequestDispatcher("/result.jsp").forward(request, response);
            return;
        }
    }

    boolean isDate(String s) { // 检测一个字符串是否能转换成日期
        SimpleDateFormat df=new  SimpleDateFormat("yyyy-MM-dd");
        df.setLenient(false); // 设置为严格验证
        try {
            df.parse(s);
            return true;
        } catch (Exception e) {
            return false;
        }
    }
}
```

4．编写 UserService 类

UserService 的具体代码如下：

UserService.java
```java
package com.service;
public class UserService {
    private UserDao userDao = new UserDaoImpl();
    /**
     * 用户登录
     * @param user：用户登录信息
     * @return 返回 user 不为空表示成功登录，否则用户名或者密码错误
     */
    public User login(User user) throws SQLException {
        return userDao.selectByNameAndPassword(user);
    }

    // 查询用户信息
    public List<User> queryUsers(User user) {
        return userDao.select(user);
    }
}
```

5．编写 UserDao 类

在 UserDao 中使用 3.1.3 节中的实体类、DBHelper 类和数据库表，具体代码如下：

```java
// UserDao 类 (UserDao.java)
package com.dao;
public interface UserDao {
    int insert(User user);
    int delete(User user);
    int update(User user);
    List<User>select(User user);
    User selectById(String id);
```

```java
    boolean isExist(User user);
    User selectByNameAndPassword(User user) throws SQLException;
}
//UserDaoImpl 类 (UserDaoImpl.java)
package com.dao.impl;
public class UserDaoImpl implements UserDao {
// 将 user 插入数据库
    @Override
    public int insert(User user) {
        String sql = "insert into t_user(name, password, sex, birthday, interest) values(?, ?, ?, ?, ?)";
        Object[] objs = {user.getName(), user.getPassword(), user.getSex(), user.getBirthday(), user.getInterest()};
        return DBHelper.ExecSql (sql, objs);
    }

// 将 user 从数据库中删除
    @Override
    public int delete(User user) {
        String sql = "delete from t_user where id = ?";
        return DBHelper.ExecSql (sql, new Object[]{user.getId()});
    }

// 修改 user 对象，并持久化到数据库
    @Override
    public int update(User user) {
        return 0;
    }

// 查询 user 对象
    @Override
    public List<User> select(User user) {
        List<User> users = new ArrayList<User>();
        ResultSet rs = null;
        if(null == user) { // User 为空则查询全部
            String sql = "select * from t_user";
            rs = DBHelper.ExecSql (sql);
        } else { // User 不为空，则查询 User
            String sql = "select * from t_user where name = ?";
            Object[] objs = {user.getName()};
            rs = DBHelper.ExecSql (sql, objs);
        }

        try {
            while(rs.next()) { // 遍历结果集
                User u = new User(rs.getInt(1), rs.getString(2), rs.getString(3),
                        rs.getString(4), rs.getString(5), rs.getString(6));
                users.add(u);
            }
```

```java
        } catch (SQLException e) {
            e.printStackTrace();
        }
        return users;
    }

    // 查询 id 对应的数据库记录
    @Override
    public User selectById(String id) {
        return null;  // 尚未实现
    }

    // 查询 user 对象是否存在于数据库中
    @Override
    public boolean isExist(User user) {
        // 尚未实现
    }

    // 通过用户名和密码查询 user 对象
    @Override
    public User selectByNameAndPassword(User user) throws SQLException {
        User u = null;
        String sql = "select * from t_user where name = ?and password = ?";
        Object[] objs = {user.getName(), user.getPassword()};
        ResultSet rs = DBHelper.ExecSql (sql, objs);
        if(rs.next()) { // 遍历结果集
            u = new User(rs.getInt(1), rs.getString(2), rs.getString(3),
                rs.getString(4), rs.getString(5), rs.getString(6));
        }
        return u;
    }
}
```

6. 新建并编辑显示登录信息页面

新建 result.jsp，并在页面中使用 EL 表达式显示用户信息，具体代码如下：

result.jsp

```jsp
<%@ page language="java" contentType="text/html; charset=UTF-8" pageEncoding="UTF-8"%>
<!DOCTYPE html PUBLIC "-//W3C//DTD HTML 4.01 Transitional//EN" "http://www.w3.org/TR/html4/loose.dtd">
<html>
<head>
  <meta http-equiv="Content-Type" content="text/html; charset=UTF-8">
  <link rel="stylesheet" type="text/css" href="css/style.css">
  <title> 结果页面 </title>
</head>
<body>
  用户 ${user.name} 登录成功 <br/>
  提交的用户名是：${user.name}<br/>
```

提交的密码是：${user.password}
　　<div class = "error">${error}</div>
</body>
</html>

7．运行项目

点击工具栏上的"Run"按钮（快捷键 Ctrl+F11），在弹出的窗口中选择"Run On Server"，启动完成后在浏览器中输入"http://localhost:8080/servlet"打开页面，效果如图 4.4 所示。

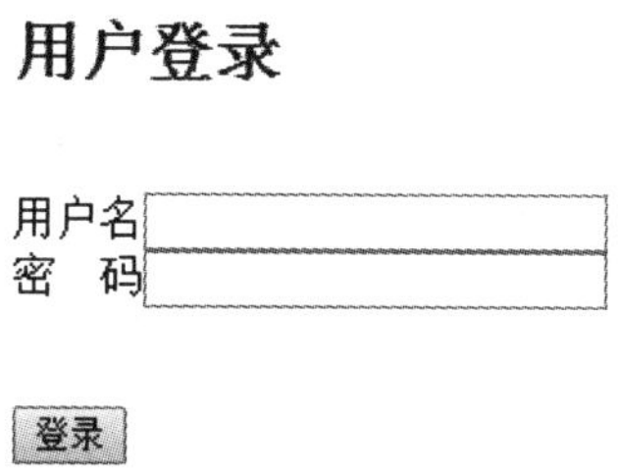

图 4.4　登录页面

如果用户填写的登录信息没有验证通过，程序会为用户响应失败信息，登录失败页面效果如图 4.5 所示。

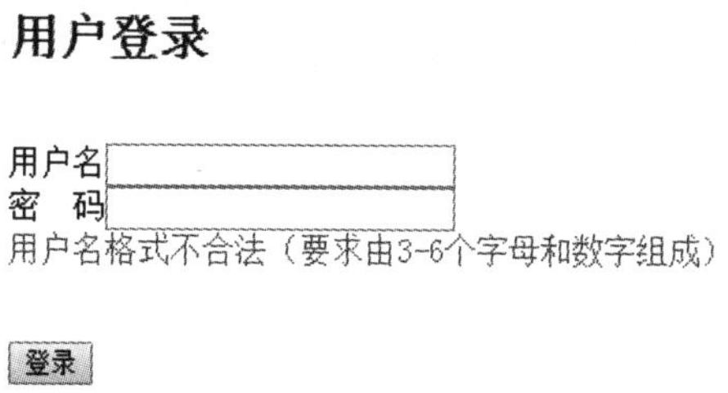

图 4.5　登录失败页面效果

如果登录用户名和密码不匹配，程序会为用户提供登录失败信息，具体效果如图 4.6 所示。

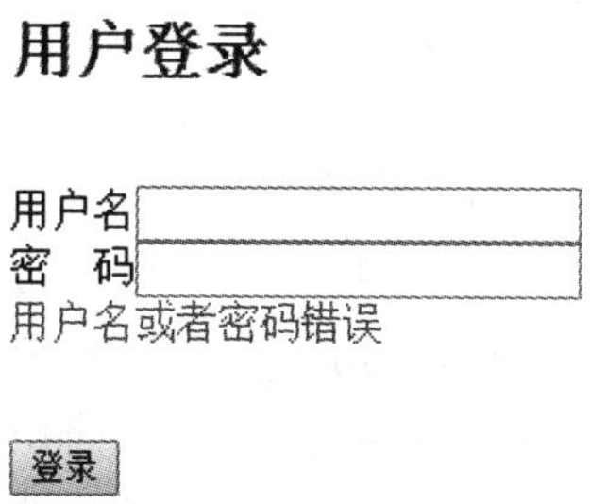

图 4.6　用户名或密码错误页面效果

如果登录成功,会显示登录成功页面,具体效果如图 4.7 所示。

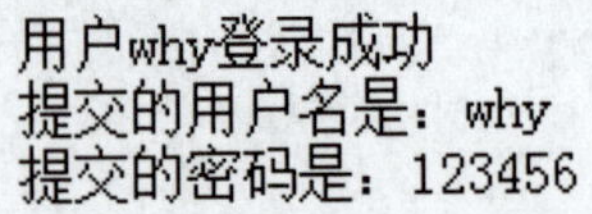

用户why登录成功
提交的用户名是:why
提交的密码是:123456

图 4.7　登录成功页面效果

【实践训练】

【任务实施】中的代码通过 Session 机制实现会话跟踪,现要求改用 Cookie 机制实现会话跟踪。

任务 4.2　用户管理

【任务分析】

管理员登录系统成功之后,跳转到用户列表页面,查看数据库中的用户信息。在该页面对每条用户信息进行修改,修改成功后回到用户列表页面显示新的数据。

【相关知识】

4.2.1　JSTL 标签

在项目 3 中已经学习了 EL 表达式,在 JSP 中,使用 EL 表达式能够使代码变得简洁,同时很方便地获取 Java 对象及属性,即使为空也不会抛出 NullPointerException 异常。但是 EL 表达式也有其局限性,比如不能进行条件判断、遍历集合等,此时就需要 JSTL 的支持。将 EL 表达式和 JSTL 标签配合使用,基本就能够满足日常开发的需要。

(1)JSTL 标签概述

JSTL(JSP Standard Tag Library)是指 JSP 标准标签库,而标签是 JSP 作为 MVC 模式中视图层的一种解决方案。JSTL 标签是由 JCP(Java Community Process)所制定的标准规范,主要为 Java Web 开发人员提供一个标准通用的标签函数库。

JSTL 标签包括 core、fmt、sql 和 XML 等 4 个标签库,以及 1 个 fn 方法库,其中 core 标签库是 JSTL 的核心标签库,主要实现数据的输出、集合的遍历、字符串处理等功能;fmt 标签库主要完成数据格式转换的功能,fmt 的英文全称为 format;sql 标签库提供了对数据库相关操作的支持,sql 标签库操作数据库比直接使用脚本操作数据库简单；XML 标签库提供了对 XML 文件解析的支持；fn 是 function 的简写，fn 方法库主要提供一些方法，以便程序调用。本节将详细讲解 JSTL 的 core 标签库，读者可以查阅资料了解其他标签库和 fn 方法库的用法。

（2）JSTL 标签优势

在之前的项目中，实现在 JSP 中显示数据的方式是使用脚本（即使用"<% %>"嵌入 Java 代码）。使用这种方式有很多弊端：首先，脚本和 HTML 代码混在一起，不利于代码的阅读和维护；其次，脚本和 HTML 代码混在一起也不利于代码的重用；最后，使用脚本实现复杂的功能比较困难。

而通过使用 EL 表达式和 JSTL 标签完成数据显示就会避免这些问题，JSTL 标签严格遵循 XML 标签的语法，使代码便于阅读和维护。JSTL 标签可以实现代码的可重用，同时使用简单的标签就能够实现比较复杂的功能。因此，在 JSP 中，不推荐使用脚本，推荐使用 EL 表达式和 JSTL 标签方式完成数据显示功能。

（3）JSTL core 标签库

从 Java EE 5 开始，内置了对 JSTL 的支持，在 JSP 中可以直接使用。Java EE 5 以前的版本需要引入 JSTL 相关 jar 包（jstl.jar，standard.jar）和 tld 文件。JSTL 的 jar 包可从 apache.org 网站下载，将 jar 包放入 Web 应用的 \WEB-INF\lib 下面，将相应的 tld 文件放到 \WEB-INF\tld 下面。

在 JSP 中使用 JSTL 标签之前，需要使用 taglib 行为标签导入 JSTL 库，下面的代码导入了 JSTL 的 core 标签库，core 标签库的 uri 规定为 http://java.sun.com/jsp/jstl/core，prefix 属性为标签库指定前缀，其值可以任意指定，一般为 c，具体代码如下：

<%@ taglib prefix="c" uri="http://java.sun.com/jsp/jstl/core"%>

JSTL core 标签库中的标签大致可以分为 4 类，分别为表达式控制标签、流程控制标签、循环标签和 URL 操作标签。下面将分别对这 4 类标签进行学习。

① 表达式控制标签

表达式控制标签主要包括 <c:out> 标签、<c:set> 标签、<c:remove> 标签和 <c:catch> 标签，下面将分别对这 4 类标签进行学习。

● <c:out> 标签

JSTL 中使用 <c:out> 标签来输出数据，<c:out> 标签有 3 个属性，具体如表 4.3 所示。

表 4.3 <c:out> 标签的属性

属性名	属性描述
value	指定输出的值，可以为 EL 表达式
default	若 value 属性为空，会输出 default 的值
escapeXml	若指定为 true，会对输出内容进行 XML 编码（只会对特殊字符如 &、>、< 等编码），默认值为 true

下面通过一个 JSTL 输出数据的例子，来了解 <c:out> 标签的常用属性。
out.jsp

```
<!DOCTYPE html PUBLIC "-//W3C//DTD HTML 4.01 Transitional//EN" "http://www.w3.org/TR/html4/loose.dtd">
<%@ page language="java" import="java.util.*" pageEncoding="utf-8" import="com.service.UserService" import="com.entity.User"%>
<%@ taglib prefix="c" uri="http://java.sun.com/jsp/jstl/core"%>
<html>
<head>
<title>c:out 标签 </title>
</head>
<body>
    flag 参数 ( 使用 jstl 方式获取 )：<c:out value="${param.flag}" default=" 未指定 flag 参数 "></c:out><br/>
    flag 参数 ( 使用 el 表达式方式获取 )：${param.flag}
</body>
</html>
```

输入网址 http://localhost:8080/servlet/out.jsp 后，显示的结果如图 4.8 所示。

```
flag参数(使用jstl方式获取)：未指定flag参数
flag参数(使用el表达式方式获取)：
```

图 4.8 <c:out> 标签属性示例测试结果（1）

输入网址 http://localhost:8080/servlet/out.jsp?flag=why 后，显示的结果如图 4.9 所示。

```
flag参数(使用jstl方式获取)：why
flag参数(使用el表达式方式获取)：why
```

图 4.9 <c:out> 标签属性示例测试结果（2）

从上述示例可以看出，当 flag 不为空时，<c:out> 标签和 EL 表达式输出的内容是相同的，但当 flag 为空时，<c:out> 标签会输出设定好的默认值，而 EL 表达式不会输出任何信息。而且当 <c:out> 标签的 escapeXml 属性设置为 true 时，<c:out value='${"<>"}'></c:out> 会输出 "<>"，而 ${"<>"} 会输出 "<>"，而且 <c:out> 标签会根据不同的环境来决定输出的格式，因此 <c:out> 标签和 EL 表达式之间还是有区别的。

- <c:set> 标签

JSTL 中使用 <c:set> 标签来设置 Java 对象属性值，<c:set> 标签有 5 个属性，具体如表 4.4 所示。

表 4.4 <c:set> 标签的属性

属性名	属性描述
var	需要设置新值的对象名，如果该对象不存在则生成，若存在则修改。var 只能设置 Integer、Double、Float、String 等类型数据，不能操作 Java Bean 或 Map 的复杂数据类型
value	需要设置的新值，可以为 EL 表达式

续表

属性名	属性描述
scope	指定对象的范围，可以为 Session、request、page、application 等，默认为 page
target	用来操作 Java Bean 或 Map 等数据类型，与 var 功能互补，两者不能同时使用，target 只接受 EL 表达式，一般与 property 配合使用，若 target 为 Java Bean，则 property 为 Java Bean 的一个属性；若 target 为 Map，则 property 为 Map 的一个 key
property	指定要修改对象的属性或键值，一般与 target 配合使用

下面通过一个 JSTL 设置数据的例子，来了解 <c:set> 标签的常用属性。

set.jsp

```
<!DOCTYPE html PUBLIC "-//W3C//DTD HTML 4.01 Transitional//EN" "http://www.w3.org/TR/html4/loose.dtd">
<%@ page language="java" import="java.util.*" pageEncoding="utf-8" import="com.service.UserService" import="com.entity.User"%>
<%@ taglib prefix="c" uri="http://java.sun.com/jsp/jstl/core"%>
<html>
<head>
<title>c:set 标签 </title>
</head>
<body>
   <c:set var="totalCount" value="${totalCount+1}" scope="application"></c:set>
   <c:set var="count" value="${count+1}" scope="session"></c:set>
   本网站总访问次数：${totalCount}<br/>
   其中您的访问次数：${count}<br/>
</body>
</html>
```

在上面的例子中，利用 <c:set> 标签的 scope 属性，记录了网站的总访问次数和当前用户的访问次数。

<c:set> 标签也支持标签体，value 值可以写在 <c:set> 标签体内，以下两种写法是等价的。

```
<c:set var="name" value="why"></c:set>
<c:set var="name">why</c:set>
```

在使用 target 属性时，要注意 target 属性只能修改已经存在的 Java Bean 或者 Map 的内容，而不能创建 Java Bean 或者 Map，因此 target 为 null 时会抛出异常，因此在使用前可以先判断 target 对象是否为 null。另外，如果 target 为 Java Bean，property 属性必须存在，否则也会抛出异常。

● <c:remove> 标签

JSTL 中使用 <c:remove> 标签能够移除所有类型的数据，<c:remove> 标签有两个属性，具体如表 4.5 所示。

表 4.5 <c:remove> 标签的属性

属性名	属性描述
var	需要移除的对象名，var 只能接受字符串而不能接受 EL 表达式
scope	指定移除对象的范围，可以为 session、request、page、application 等

下面通过一个 JSTL 移除数据的例子，来了解 <c:remove> 标签的常用属性。

remove.jsp
```
<!DOCTYPE html PUBLIC "-//W3C//DTD HTML 4.01 Transitional//EN" "http://www.w3.org/TR/html4/loose.dtd">
<%@ page language="java" import="java.util.*" pageEncoding="utf-8" import="com.service.UserService" import="com.entity.User"%>
<%@ taglib prefix="c" uri="http://java.sun.com/jsp/jstl/core"%>
<%
    request.setAttribute("list", new ArrayList());
%>
<html>
<head>
  <title>c:remove 标签 </title>
</head>
<body>
  <c:remove var="list"/>
  ${list == null ? 'list 已经被删除 ' : 'list 没有被删除 '}
</body>
</html>
```

上述示例将会输出"list 已经被删除"。

<c:remove> 标签可以从 Page、Request、Session、Application 等范围内删除任何类型的数据，即使 var 指定的对象不存在，<c:remove> 标签也不会抛出异常。

- <c:catch> 标签

JSTL 中使用 <c:catch> 标签捕获 JSP 运行时抛出的异常，<c:catch> 标签只有 1 个属性 var，用于指定存储异常信息的变量名。下面通过一个 JSTL 抛出异常的例子，来了解 <c:catch> 标签的 var 属性。

catch.jsp
```
<!DOCTYPE html PUBLIC "-//W3C//DTD HTML 4.01 Transitional//EN" "http://www.w3.org/TR/html4/loose.dtd">
<%@ page language="java" import="java.util.*" pageEncoding="utf-8" import="com.service.UserService" import="com.entity.User"%>
<%@ taglib prefix="c" uri="http://java.sun.com/jsp/jstl/core"%>
<html>
<head>
<title>c:catch 标签 </title>
</head>
<body>
  <c:catch var="ex">
    <c:set target="user" property="name" value="why"></c:set>
  </c:catch>
```

程序抛出了异常 ${ex.class.name}，原因：${ex.message}
     </body>
     </html>

上述示例中，将 <c:set> 标签的 target 属性指定了字符串，这样代码会抛出异常，程序使用 <c:catch> 标签将异常信息存入 ex 对象中，并打印了出来。

②流程控制标签

流程控制标签主要包括 <c:if> 标签、<c:choose> 标签、<c:when> 标签和 <c:otherwise> 标签，下面将分别对这 4 类标签进行学习。

● <c:if> 标签

JSTL 中使用 <c:if> 标签实现类似 Java 中 if 判断的功能，下面通过一个示例来了解 <c:if> 标签的用法。

if.jsp

```
<!DOCTYPE html PUBLIC "-//W3C//DTD HTML 4.01 Transitional//EN" "http://www.w3.org/TR/html4/loose.dtd">
<%@ page language="java" import="java.util.*" pageEncoding="utf-8" import="com.service.UserService" import="com.entity.User"%>
<%@ taglib prefix="c" uri="http://java.sun.com/jsp/jstl/core"%>
<html>
<head>
<title>c:if 标签 </title>
</head>
<body>
   <c:if test="${param.flag == 'why'}">
      您输入的 flag 的值为 why
   </c:if>
   <c:if test="${param.flag == null}">
      您没有输入 flag
   </c:if>
</body>
</html>
```

在输入 http://localhost:8080/servlet/out.jsp?flag=why 后，会输出"您输入的 flag 的值为 why"，在输入 http://localhost:8080/servlet/out.jsp 后，会输出"您没有输入 flag"。

通过上述示例可以看出，只有 <c:if> 标签的 test 属性值为 boolean 类型的 true 或为字符串类型的 true（大小写不敏感）时，if 标签体才会执行。

● <c:choose> 标签、<c:when> 标签、<c:otherwise> 标签

<c:if> 标签没有 else 的功能，若要想实现类似于 Java 中的 if…else 的流程，需要使用 <c:choose>、<c:when>、<c:otherwise> 等 3 个标签，下面通过一个示例进行演示。

ifElse.jsp

```
<!DOCTYPE html PUBLIC "-//W3C//DTD HTML 4.01 Transitional//EN" "http://www.w3.org/TR/html4/loose.dtd">
<%@ page language="java" import="java.util.*" pageEncoding="utf-8" import="com.service.UserService" import="com.entity.User"%>
<%@ taglib prefix="c" uri="http://java.sun.com/jsp/jstl/core"%>
<html>
<head>
```

```
      <title>if...else 流程实现 </title>
    </head>
    <body>
      <c:choose>
        <c:when test="${param.flag == null}">
           您没有输入 flag 的值
        </c:when>
        <c:otherwise>
           您输入的 flag 的值为 ${param.flag}
        </c:otherwise>
      </c:choose>
    </body>
</html>
```

在输入 http://localhost:8080/servlet/out.jsp?flag=why 后，会输出"您输入的 flag 的值为 why"，在输入 http://localhost:8080/servlet/out.jsp 后，会输出"您没有输入 flag 的值"。

③循环标签

循环标签主要包括 <c:forEach> 标签和 <c:forTokens> 标签，下面将分别对这两类标签进行学习。

- <c:forEach> 标签

JSTL 中使用 <c:forEach> 标签实现类似于 Java 中的 while 和 for 的功能，<c:forEach> 标签有 6 个属性，具体如表 4.6 所示。

表 4.6 <c:forEach> 标签的属性

属性名	属性描述
var	用于存放当前指定的对象
items	被迭代的集合对象，可以是数组、集合、Map 和 String 等类型的对象，也可以是 EL 表达式
varStatus	用于存放当前指定对象的相关信息
begin	开始迭代的位置，默认值为 0
end	结束迭代的位置，默认值为最后一个对象的位置
step	每次迭代的间隔数，默认值为 1

下面通过一个示例来了解 <c:forEach> 标签的用法，该示例打印 1～100 内的所有偶数。

forEach.jsp
```
<!DOCTYPE html PUBLIC "-//W3C//DTD HTML 4.01 Transitional//EN" "http://www.w3.org/TR/html4/loose.dtd">
<%@ page language="java" import="java.util.*" pageEncoding="utf-8" import="com.service.UserService" import="com.entity.User"%>
<%@ taglib prefix="c" uri="http://java.sun.com/jsp/jstl/core"%>
<html>
<head>
<title>c:forEach 标签 </title>
```

```jsp
</head>
<body>
    <c:forEach var="num" begin="2" end="100" step="2">
        ${num}
    </c:forEach>
</body>
</html>
```

<c:forEach> 标签不仅可以当做循环使用，还可以遍历集合类型的对象，下面将通过一个示例说明如何使用 <c:forEach> 标签遍历 List 集合对象。

list.jsp

```jsp
<!DOCTYPE html PUBLIC "-//W3C//DTD HTML 4.01 Transitional//EN" "http://www.w3.org/TR/html4/loose.dtd">
<%@ page language="java" import="java.util.*" pageEncoding="utf-8" import="com.service.UserService" import="com.entity.User"%>
<%@ taglib prefix="c" uri="http://java.sun.com/jsp/jstl/core"%>
<%
    List<User> users = new ArrayList<User>();
    User u1 = new User("why1", "111111");
    User u2 = new User("why2", "222222");
    users.add(u1);
    users.add(u2);
    request.setAttribute("users", users);
%>
<html>
<head>
<title>c:forEach 标签 </title>
<link rel="stylesheet" type="text/css" href="css/style.css">
</head>
<body>
    <table class="gridtable" width="40%">
        <tr>
            <th> 用户名 </th>
            <th> 密码 </th>
        </tr>
        <c:forEach items="${users}" var="user">
            <tr>
                <td>${user.name}</td>
                <td>${user.password}</td>
            </tr>
        </c:forEach>
    </table>
</body>
</html>
```

上述程序运行后的结果如图 4.10 所示。

用户名	密码
why1	111111
why2	222222

图 4.10 <c:forEach> 标签遍历 List 结果

在上述代码中，items 属性使用 EL 表达式传入 List 对象，代码并没有告诉 JSTL 标签 users 是什么类型，JSTL 在运行时会通过反射机制获取数据类型及相关属性值，在定义 User 类时，一定要为属性设定 getter 和 setter 方法，这样 JSTL 才能够获取和设置属性值。

<c:forEach> 标签还可以遍历 Map 类型的对象，下面将通过一个示例说明如何使用 <c:forEach> 标签遍历 Map 集合对象。

map.jsp
```
<!DOCTYPE html PUBLIC "-//W3C//DTD HTML 4.01 Transitional//EN" "http://www.w3.org/TR/html4/loose.dtd">
<%@ page language="java" import="java.util.*" pageEncoding="utf-8" import="com.service.UserService" import="com.entity.User"%>
<%@ taglib prefix="c" uri="http://java.sun.com/jsp/jstl/core"%>
<html>
<head>
<title>c:forEach 标签 </title>
<link rel="stylesheet" type="text/css" href="css/style.css">
</head>
<body>
   <table class="gridtable" width="50%">
     <tr>
       <th>name</th>
       <th>value</th>
     </tr>
     <c:forEach items="${header}" var="item">
       <tr>
         <td>${item.key}</td>
         <td>${item.value}</td>
       </tr>
     </c:forEach>
   </table>
</body>
</html>
```

上述程序运行后的结果如图 4.11 所示。

name	value
cookie	JSESSIONID=6B1E3D9FDBE3890B3A7904A240EF1CC5
cache-control	max-age=0
connection	keep-alive
accept-language	zh-CN,zh;q=0.8
host	localhost:8080
accept	text/html,application/xhtml+xml,application/xml;q=0.9,image/webp,*/*;q=0.8
user-agent	Mozilla/5.0 (Windows NT 6.1) AppleWebKit/537.36 (KHTML, like Gecko) Chrome/45.0.2454.101 Safari/537.36
accept-encoding	gzip, deflate, sdch
upgrade-insecure-requests	1

图 4.11 <c:forEach> 标签遍历 Map 结果

由上述代码可知，遍历 Map 和遍历 List 是有区别的，因为 Map 中是通过键值对来存储信息的，因此在获取到头信息中的每个键值对 item 后，要使用 ${item.key} 获取键值，使用 ${item.value} 获取键对应的实际值。

在前面的例子中，只是简单地遍历了集合中的对象，但并不知道该对象是第几个，如果想要知道当前正在遍历对象的相关信息，必须使用 <c:forEach> 标签的 varStatus 属性，varStatus 属性存储当前正在遍历的对象的信息，varStatus 中常用的属性如表 4.7 所示。

表 4.7　varStatus 中常用的属性

属性名	属性描述
index	返回当前对象是第几个对象，从 0 开始计数
count	返回已经遍历了多少个对象，从 1 开始计数
first	返回当前对象是否是第一个对象
last	返回当前对象是否是最后一个对象
current	返回当前对象
begin	返回 <c:forEach> 标签 begin 的属性值
end	返回 <c:forEach> 标签 end 的属性值
step	返回 <c:forEach> 标签 step 的属性值

下面通过一个示例来了解 varStatus 属性的用法。

varStatus.jsp
<!DOCTYPE html PUBLIC "-//W3C//DTD HTML 4.01 Transitional//EN" "http://www.w3.org/TR/html4/loose.dtd">
<%@ page language="java" import="java.util.*" pageEncoding="utf-8" import="com.service.UserService" import="com.entity.User"%>
<%@ taglib prefix="c" uri="http://java.sun.com/jsp/jstl/core"%>

```
<%
    List<User> users = new ArrayList<User>();
    User u1 = new User("why1", "111111");
    User u2 = new User("why2", "222222");
    User u3 = new User("why3", "333333");
    User u4 = new User("why4", "444444");
    users.add(u1);
    users.add(u2);
    users.add(u3);
    users.add(u4);
    request.setAttribute("users", users);
%>
<html>
<head>
<title>c:forEach 标签的 varStatus 属性 </title>
<link rel="stylesheet" type="text/css" href="css/style.css">
</head>
<body>
    <table class="gridtable" width="40%">
      <tr>
         <th> 用户名 </th>
         <th> 密码 </th>
      </tr>
      <c:forEach items="${users}" var="user" varStatus="varStatus">
         <tr bgcolor="${varStatus.index % 2 == 1 ? '#FFFFFF' : '#EFEFEF'}">
            <td>${user.name}</td>
            <td>${user.password}</td>
         </tr>
      </c:forEach>
    </table>
</body>
</html>
```

上述示例通过 varStatus 的 index 属性判断当前行是奇数行还是偶数行，从而实现了表格的奇数行增添背景色，而偶数行不变色的效果，程序的运行结果如图 4.12 所示。

用户名	密码
why1	111111
why2	222222
why3	333333
why4	444444

图 4.12　varStatus 属性的示例运行结果

● <c:forTokens> 标签

<c:forTokens> 标签与 <c:forEach> 标签类似，都有 var、begin、end、step、items

和 varStatus 属性，不同的是 <c:forEach> 标签的 items 属性存放的是集合类型或者数组类型的对象，而 <c:forTokens> 标签的 items 属性存放的是字符串，且这个字符串会被 delims 属性中的分隔符分割成多个字符串。下面通过一个示例来了解 <c:forTokens> 标签的用法。

forTokens.jsp

```jsp
<!DOCTYPE html PUBLIC "-//W3C//DTD HTML 4.01 Transitional//EN" "http://www.w3.org/TR/html4/loose.dtd">
<%@ page language="java" import="java.util.*" pageEncoding="utf-8" import="com.service.UserService" import="com.entity.User"%>
<%@ taglib prefix="c" uri="http://java.sun.com/jsp/jstl/core"%>
<%
    List<User> users = new ArrayList<User>();
    User u1 = newUser("why1", "111111");
    User u2 = newUser("why2", "222222");
    User u3 = newUser("why3", "333333");
    User u4 = newUser("why4", "444444");
    users.add(u1);
    users.add(u2);
    users.add(u3);
    users.add(u4);
    request.setAttribute("users", users);
%>
<html>
<head>
<title>c:forTokens 标签 </title>
<link rel="stylesheet" type="text/css" href="css/style.css">
</head>
<body>
    <table class="gridtable" width="40%">
        <tr>
            <th>index</th>
            <th>name</th>
        </tr>
        <c:forTokens items="why1,why2,why3,why4" delims="," var="item" varStatus="varStatus">
            <tr>
                <td>${varStatus.index}</td>
                <td>${item}</td>
            </tr>
        </c:forTokens>
    </table>
</body>
</html>
```

程序的运行结果如图 4.13 所示。

index	name
0	why1
1	why2
2	why3
3	why4

图 4.13　<c:forTokens> 标签示例运行结果

④ URL 操作标签

URL 操作标签主要包括 <c:import> 标签、<c:url> 标签、<c:redirect> 标签和 <c:param> 标签，下面将分别对这 4 类标签进行学习。

● <c:import> 标签

<c:import> 标签主要用于引入目标网页，可以是网络资源，也可以是本项目内的资源，<c:import> 标签有 6 个属性，具体如表 4.8 所示。

表 4.8　<c:import> 标签的属性

属性名	属性描述
url	指定目标网页的 url
charEncoding	指定目标网页的编码格式，默认为 ISO-8859-1
var	指定目标网址的存储变量名，该变量为 String 类型
varReader	指定目标网址的存储变量名，该变量为 java.io.Reader 类型，不能与 var 属性或者 scope 属性共存
context	指定目标网页的 contextPath，url 与 context 必须以 "/" 开始，目标网页必须是本服务器内的，如 <c:import url="/login.jsp" context="/servlet"> 会请求 /servlet/login.jsp，若为本 Web 应用内的资源，context 可省略
scope	指定 var 变量的范围

下面通过一个示例来了解 <c:import> 标签的用法。

import.jsp

```
<!DOCTYPE html PUBLIC "-//W3C//DTD HTML 4.01 Transitional//EN" "http://www.w3.org/TR/html4/loose.dtd">
<%@ page language="java" import="java.util.*" pageEncoding="utf-8" import="com.service.UserService" import="com.entity.User"%>
<%@ taglib prefix="c" uri="http://java.sun.com/jsp/jstl/core"%>
<html>
<head>
<title>c:import 标签 </title>
</head>
<body>
    <c:import url="http://www.baidu.com" var="baidu" charEncoding="utf-8" scope="request"></
```

```
   c:import>
      百度的源码为: <br/><br/>
      <c:out value="${baidu}" escapeXml="true"></c:out>
</body>
</html>
```
程序的运行结果如图 4.14 所示。

```
百度的源码为:

<!DOCTYPE html> <!--STATUS OK--><html> <head><meta http-equiv=content-type
content=text/html;charset=utf-8><meta http-equiv=X-UA-Compatible content=IE=Edge><meta
content=always name=referrer><link rel=stylesheet type=text/css
href=http://s1.bdstatic.com/r/www/cache/bdorz/baidu.min.css><title>百度一下,你就知道
</title></head> <body link=#0000cc> <div id=wrapper> <div id=head> <div class=head_wrapper>
<div class=s_form> <div class=s_form_wrapper> <div id=lg> <img hidefocus=true
src=//www.baidu.com/img/bd_logo1.png width=270 height=129> </div> <form id=form name=f
action=//www.baidu.com/s class=fm> <input type=hidden name=bdorz_come value=1> <input
type=hidden name=ie value=utf-8> <input type=hidden name=f value=8> <input type=hidden
name=rsv_bp value=1> <input type=hidden name=rsv_idx value=1> <input type=hidden name=tn
value=baidu><span class="bg s_ipt_wr"><input id=kw name=wd class=s_ipt value maxlength=255
autocomplete=off autofocus></span> <span class="bg s_btn_wr"><input type=submit id=su value=
百度一下 class="bg s_btn"></span> </form> </div> </div> <div id=u1> <a
href=http://news.baidu.com name=tj_trnews class=mnav>新闻</a> <a href=http://www.hao123.com
name=tj_trhao123 class=mnav>hao123</a> <a href=http://map.baidu.com name=tj_trmap
class=mnav>地图</a> <a href=http://v.baidu.com name=tj_trvideo class=mnav>视频</a> <a
href=http://tieba.baidu.com name=tj_trtieba class=mnav>贴吧</a> <noscript> <a
href=http://www.baidu.com/bdorz/login.gif?
login&amp;tpl=mn&amp;u=http%3A%2F%2Fwww.baidu.com%2f%3fbdorz_come%3d1 name=tj_login
class=lb>登录</a> </noscript> <script>document.write('<a
href="http://www.baidu.com/bdorz/login.gif?login&tpl=mn&u='+
encodeURIComponent(window.location.href+ (window.location.search === "" ? "?" : "&")+
"bdorz_come=1")+ '" name="tj_login" class="lb">登录</a>');</script> <a
href=//www.baidu.com/more/ name=tj_briicon class=bri style="display: block;">更多产品</a>
</div> </div> </div> <div id=ftCon> <div id=ftConw> <p id=lh> <a
href=http://home.baidu.com>关于百度</a> <a href=http://ir.baidu.com>About Baidu</a> </p> <p
id=cp>&copy;2016 Baidu <a href=http://www.baidu.com/duty/>使用百度前必读
</a>  <a href=http://jianyi.baidu.com/ class=cp-feedback>意见反馈</a> 京ICP证
030173号  <img src=//www.baidu.com/img/gs.gif> </p> </div> </div> </div> </body>
</html>
```

图 4.14 <c:import> 标签示例的运行结果

- <c:url> 标签

当客户端浏览器不支持 Cookie 或者 Cookie 被禁用后,可以采用 response.encodeURL() 方法对 URL 地址重写,使 Session 功能能够继续使用,<c:url> 标签能够实现对 URL 地址进行重写。比如 login.jsp 中存在代码 <c:url value="/images/bg.gif"/>,若客户端浏览器不支持 Cookie 功能,则在请求 login.jsp 后,该地址会被重写为如下格式。

/servlet/images/bg.gif;jsessionid=5F4771183629C9834F8382E23BE13C4C

重写后的地址自动加上了 contextPath 和 JSESSIONID,注意 URL 必须以"/"开始才会对其进行重写。

除了 value 属性外,<c:url> 标签也有 context、var 和 scope 属性,这 3 个属性用法类似于 <c:import> 标签,在此不再赘述。

- <c:redirect> 标签

<c:redirect> 标签主要用于实现重定向的功能,<c:redirect> 标签有两个属性,即 url 和 context,url 为重定向的网址,context 默认为当前的 contextPath。如果声明了 context 属性,则 context 与 url 必须以"/"开头,且 url 是相对于 context 的路径,例如

重定向到本应用的 login.jsp 的代码如下所示。

<c:redirect url="/login.jsp" context="/servlet" />

- <c:param> 标签

有时候重定向到某个网页或者引入某个网址需要指定参数，<c:param> 标签专门用于设定参数。下面通过一个例子来对 <c:param> 标签进行说明。

<c:redirect url="http://www.baidu.com">
<c:param name="name" value="why"></c:param>
  <c:param name="pwd" value="123456"></c:param>
</c:redirect>

该示例在重定向后的网址为 http://www.baidu.com? Name=why& pwd=123456。

【任务实施】

1. 用户管理功能流程分析

在编写代码之前，首先对用户管理中的查询用户、增加用户、修改用户和删除用户的流程进行分析。

①查询用户列表流程分析

在浏览器中输入地址 http://localhost:8080/servlet/，进入登录界面（login.jsp），在输入用户名和密码并验证成功后（控制器 UserServlet?flag=login），进入用户主界面（user.jsp），在用户主界面中，提供了增加、修改和删除链接。具体的主界面流程图如图 4.15 所示。

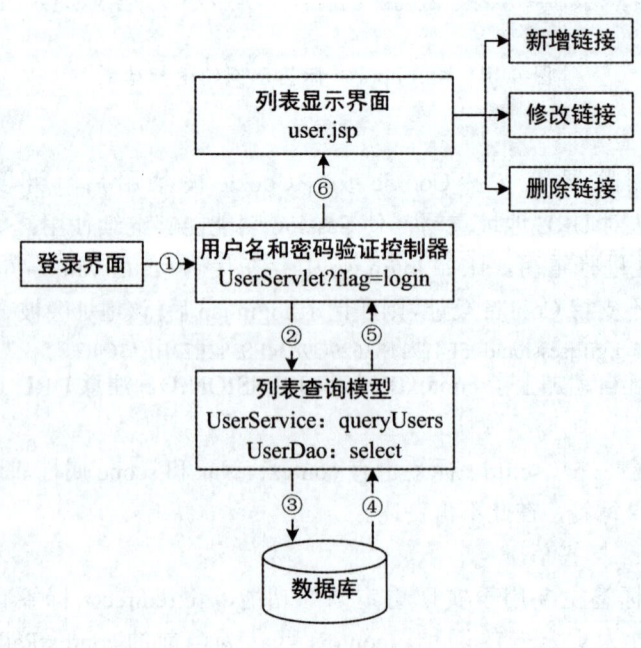

图 4.15 主界面流程图

②新增用户流程分析

在用户主界面点击"新增链接"后进入新增界面（addUser.jsp），在新增界面填写待添加用户的信息后，点击"确定"按钮，在控制器（UserServlet?flag=addUser）中验证并调用 Service 层（UserService）代码完成新增用户操作，最后进入主界面。具体新增用户流程如图 4.16 所示。

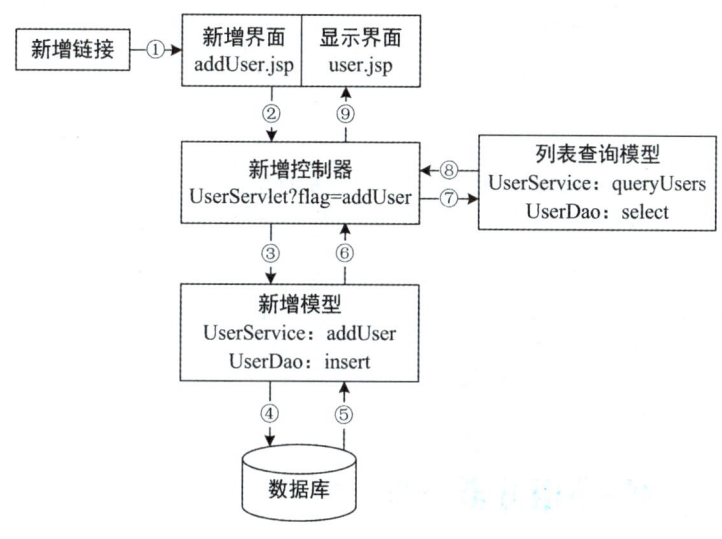

图 4.16　新增用户流程

③修改用户流程分析

在用户主界面点击"修改链接"后，进入控制器（UserServlet?flag=queryById）查询待修改的用户，然后进入修改界面并显示待修改用户信息（modify.jsp），编辑用户信息完成后，点击"确定"按钮进入控制器（UserServlet?flag=modifyUser）调用 Service 层（UserService）代码完成修改用户操作，最后进入主界面。具体修改用户流程如图 4.17 所示。

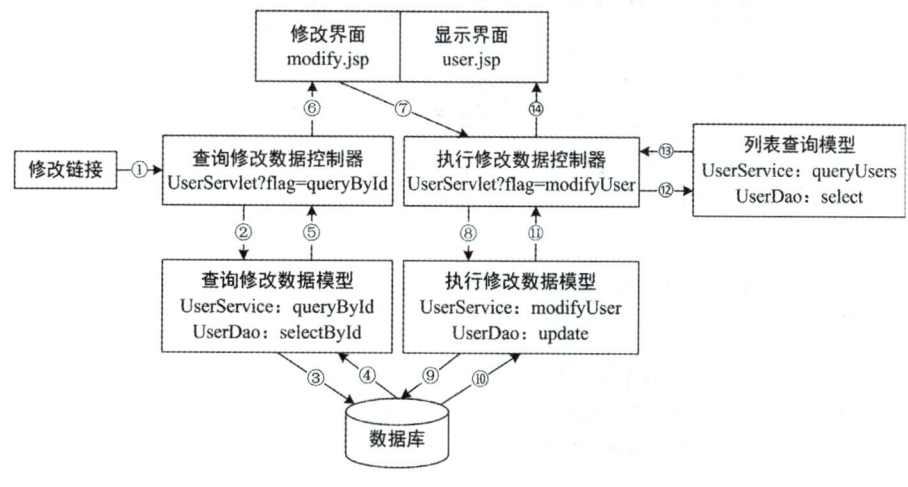

图 4.17　修改用户流程

④删除用户流程分析

在主界面点击"删除链接"后,进入控制器(UserServlet?flag=removeById)查询待删除的用户,然后进入删除界面并显示待删除用户信息(delete.jsp),点击"删除"按钮后进入控制器(UserServlet?flag=removeUser)调用 Service 层(UserService)代码完成删除用户操作,最后进入主界面。具体删除用户流程如图 4.18 所示。

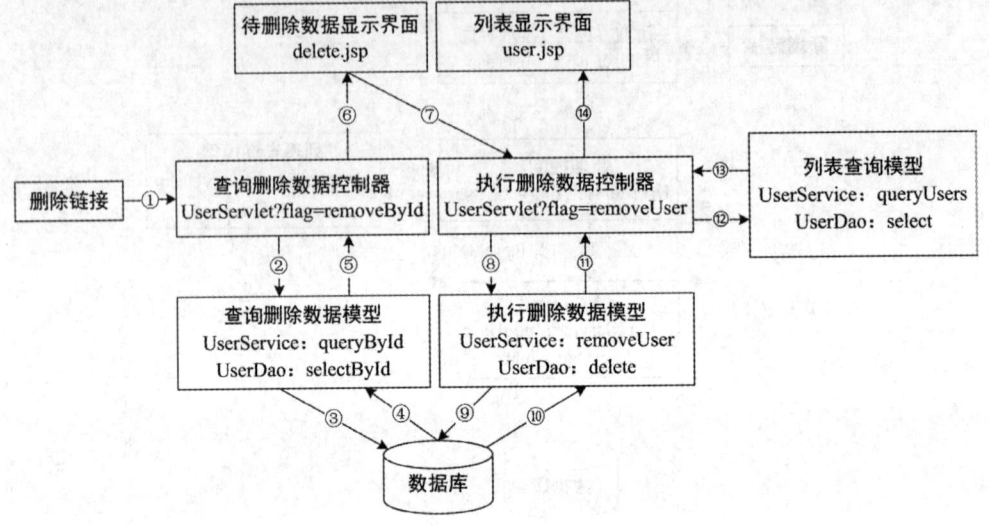

图 4.18 删除用户流程

2. 新建项目并配置 web.xml 文件

选择菜单"File"—"New"—"Dynamic Web Project"(如果没有列出,在"Other…"中查找),在弹出的窗口中填写"Project name"为"servlet",连续点击"Next"两次,在"Web Module"选项卡中勾选"Generate web.xml deployment descriptor"复选框,让项目中自动添加 web.xml 文件,最后点击"Finish"按钮创建新项目。

在 web.xml 中修改 Web 应用程序的主页设置,将主页改为 login.jsp,login.jsp 将重用 4.1.1 节中的代码,并配置 Servlet,具体代码如下:

```
web.xml
<welcome-file-list>
<welcome-file>login.jsp</welcome-file>
</welcome-file-list>
<servlet>
  <servlet-name>UserServlet</servlet-name>
  <servlet-class>com.servlet.UserServlet</servlet-class>
  <init-param>
    <param-name>encoding</param-name>
    <param-value>utf-8</param-value>
  </init-param>
  <load-on-startup>1</load-on-startup>
</servlet>
<servlet-mapping>
```

```
    <servlet-name>UserServlet</servlet-name>
    <url-pattern>/UserServlet</url-pattern>
</servlet-mapping>
```

3. 新建并编辑登录成功页面（用户列表页面）和新增界面

新建 user.jsp 页面，并输入如下代码：

user.jsp
```
<!DOCTYPE html PUBLIC "-//W3C//DTD HTML 4.01 Transitional//EN" "http://www.w3.org/TR/html4/loose.dtd">
<%@ page language="java" import="java.util.*" pageEncoding="utf-8" import="com.service.UserService" import="com.entity.User"%>
<%@ taglib prefix="c" uri="http://java.sun.com/jsp/jstl/core"%>
<html>
<head>
    <link rel="stylesheet" type="text/css" href="css/style.css">
<title> 用户列表 </title>
</head>
<body>
    欢迎：${user.name}<br/><br/>
    <table class="gridtable" width="40%">
      <tr>
        <th>id</th>
        <th>name</th>
        <th> 操作 (<a href="userAdd.jsp"> 添加 </a>)</th>
      </tr>
      <c:forEach items="${userList}" var="user">
        <tr>
          <td>${user.id}</td>
          <td>${user.name}</td>
          <td>
            <a href="UserServlet?flag=queryById&id=${user.id}"> 修改 </a>
            <a href="UserServlet?flag=removeById&id=${user.id}"> 删除 </a>
          </td>
        </tr>
      </c:forEach>
    </table>
</body>
</html>
```

新建 userAdd.jsp 页面，并输入如下代码：

userAdd.jsp
```
<%@ page language="java" contentType="text/html; charset=UTF-8" pageEncoding="UTF-8"%>
<!DOCTYPE html PUBLIC "-//W3C//DTD HTML 4.01 Transitional//EN" "http://www.w3.org/TR/html4/loose.dtd">
<html>
<head>
    <meta http-equiv="Content-Type" content="text/html; charset=UTF-8">
    <link rel="stylesheet" type="text/css" href="css/style.css">
    <title> 新增用户 </title>
</head>
```

```html
<body>
  <form action="UserServlet?flag=addUser" method="post">
    <h2> 新增用户 </h2><br>
    用户名 <input type="text" name="name"><br>
    密　码 <input type="password" name="password"><br>
    性　别 <select name="sex">
      <option value="M"> 男 </option>
      <option value="F"> 女 </option>
    </select><br>
    生　日 <input type="text" name="birthday"><br>
    爱　好
    <input type="checkbox" name="interest" value="1"> 运动
    <input type="checkbox" name="interest" value="2"> 音乐
    <input type="checkbox" name="interest" value="3"> 旅游 <br><br>
    <div class = "error">${error}</div>
    <br><br>
    <input type="submit" value=" 确定 ">
  </form>
</body>
</html>
```

4．编写 UserServlet 类

UserServlet 中会获取登录页面提交的参数，并在 UserServlet 中使用 3.1.3 节中的实体类，具体代码如下：

```java
UserServlet.java
package com.servlet;
public class UserServlet extends HttpServlet {
    private UserService service = new UserService();
    @Override
    protected void doGet(HttpServletRequest request, HttpServletResponse response) throws ServletException, IOException {
        this.doPost(request, response);
    }

    @Override
    protected void doPost(HttpServletRequest request, HttpServletResponse response) throws ServletException, IOException {
        String encoding = getInitParameter("encoding"); // 获取 web.xml 中的初始化参数
        request.setCharacterEncoding(encoding);     // 设置 request 的编码格式
        response.setCharacterEncoding(encoding);    // 设置 response 的编码格式
        String flag = request.getParameter("flag");  // 获取操作类型
        if(null != flag && flag.equals("reg")) {      // 注册
            reg(request, response);
            return;
        } else if(null != flag && flag.equals("login")) {// 登录
            try {
                login(request, response);
            } catch (SQLException e) {
                e.printStackTrace();
```

```java
                return;
            }
            return;
        } else if(null != flag && flag.equals("addUser")) { // 新增用户
            try {
                addUser(request, response);
            } catch (SQLException e) {
                e.printStackTrace();
                return;
            }
            return;
        }
    }

    /**
     * 完成登录功能
     * @param request : request 对象
     * @param response : response 对象
     */
    private void login(HttpServletRequest request, HttpServletResponse response) throws ServletException, IOException, SQLException {
        String name = request.getParameter("name");
        String password = request.getParameter("password");
        String error = "";

        Pattern p = Pattern.compile("^[A-Za-z0-9]{3,6}$");
        if(!p.matcher(name).find()) {
            error = " 用户名格式不合法（要求由 3-6 个字母和数字组成）";
            request.setAttribute("error", error);
            request.getRequestDispatcher("/login.jsp").forward(request, response);
            return;
        }
        p=Pattern.compile("^[A-Za-z0-9]{6,10}$");
        if(!p.matcher(password).find()) {
            error = " 密码格式不合法（要求由 6-10 个字母和数字组成）";
            request.setAttribute("error", error);
            request.getRequestDispatcher("/login.jsp").forward(request, response);
            return;
        }

        User user = new User(name, password);
        User u = service.login(user);
        if(null == u) {
            error = " 用户名或者密码错误 ";
            request.setAttribute("error", error);
            request.getRequestDispatcher("/login.jsp").forward(request, response);
            return;
        } else {
            HttpSession session = request.getSession(); // 获取 Session 对象
            session.setAttribute("user", u);            // 将 user 放到 Session 中
```

```java
            request.setAttribute("userList", service.queryUsers(null));// 设置列表给页面
            request.getRequestDispatcher("/user.jsp").forward(request, response);
            return;
        }
    }

    /**
     * 新增用户
     * @param request: request 对象
     * @param response: response 对象
     */
     private void addUser(HttpServletRequest request, HttpServletResponse response) throws IOException, ServletException, SQLException {
        String encoding = getInitParameter("encoding"); // 获取 web.xml 中配置的参数
        request.setCharacterEncoding(encoding);        // 为 request 指定编码格式
        response.setCharacterEncoding(encoding);       // 为 response 指定编码格式
        String name = request.getParameter("name");    // 获取页面提交的数据
        String password = request.getParameter("password");
        String sex = request.getParameter("sex");
        String birthday = request.getParameter("birthday");
        String[] interests = request.getParameterValues("interest");
        String endInterests = "";
        String error = "";
        if(null != interests) {
            for (String interest : interests) {
                endInterests += interest + ",";
            }
            endInterests = endInterests.substring(0, endInterests.length()-1);
        }

        Pattern p = Pattern.compile("^[A-Za-z0-9]{3,6}$");
        if(!p.matcher(name).find()) { // 验证用户名格式
            error = " 用户名格式不合法（要求由 3-6 个字母和数字组成）";
            request.setAttribute("error", error);
            request.getRequestDispatcher("/userAdd.jsp").forward(request, response);
            return;
        }
        p=Pattern.compile("^[A-Za-z0-9]{6,10}$");
        if(!p.matcher(password).find()) { // 验证密码格式
            error = " 密码格式不合法（要求由 6-10 个字母和数字组成）";
            request.setAttribute("error", error);
            request.getRequestDispatcher("/userAdd.jsp").forward(request, response);
            return;
        }
        if(!isDate(birthday)) { // 验证生日格式
            error = " 生日格式不合法 ";
            request.setAttribute("error", error);
            request.getRequestDispatcher("/userAdd.jsp").forward(request, response);
            return;
        }
```

```java
            User user = new User(name, password, sex, birthday, endInterests);
            UserService service = new UserService();
            int result = service.addUser(user);   // 执行添加操作
            if(0 == result){   // 添加失败
                error = " 添加失败 ";
                request.setAttribute("error", error);
                request.getRequestDispatcher("/userAdd.jsp").forward(request, response);
                return;
            } else {   // 添加成功
                request.setAttribute("userList", service.queryUsers(null));// 设置列表给页面
                request.getRequestDispatcher("/user.jsp").forward(request, response);
                return;
            }
        }
        boolean isDate(String s) { // 检测一个字符串是否能转换成日期
            SimpleDateFormat df=new  SimpleDateFormat("yyyy-MM-dd");
            df.setLenient(false); // 设置为严格验证
            try {
                df.parse(s);
                return true;
            } catch (Exception e) {
                return false;
            }
        }
    }
}
```

5. 编写 UserService 类

UserService 的具体代码如下：

UserService.java

```java
package com.service;
public class UserService {
    private UserDao userDao = new UserDaoImpl();

    /**
     * 用户登录
     * @param user：用户登录信息
     * @return：返回 user 不为空表示成功登录，否则提示用户名或者密码错误
     */
    public User login(User user) throws SQLException {
        return userDao.selectByNameAndPassword(user);
    }

    /**
     * 查询用户信息
     * @param user：参数
     * @return 用户列表
     */
    public List<User> queryUsers(User user) {
        return userDao.select(user);
```

```java
    }

    /**
     * 添加用户信息
     * @param user：参数
     * @return 受影响的行数
     */
    public int addUser(User user) {
        return userDao.insert(user);
    }
}
```

6. 编写 UserDao 类

在 UserDao 中会使用 3.1.3 节中的实体类、DBHelper 类和数据库表，具体代码如下：

```java
//UserDao(UserDao.java)
package com.dao;
public interface UserDao {
    int insert(User user);
    int delete(User user);
    int update(User user);
    List<User>select(User user);
    User selectById(String id);
    boolean isExist(User user);
    User selectByNameAndPassword(User user) throws SQLException;
}
//UserDaoImpl 类 (UserDaoImpl.java)
package com.dao.impl;
public class UserDaoImpl implements UserDao {

// 将 user 插入数据库
    @Override
    public int insert(User user) {
        String sql = "insert into t_user(name, password, sex, birthday, interest) values(?, ?, ?, ?, ?)";
        Object[] objs = {user.getName(), user.getPassword(), user.getSex(), user.getBirthday(), user.getInterest()};
        return DBHelper.ExecSql (sql, objs);
    }

// 将 user 从数据库中删除
    @Override
    public int delete(User user) {
        String sql = "delete from t_user where id = ?";
        return DBHelper.ExecSql (sql, new Object[]{user.getId()});
    }

// 修改 user 对象，并持久化到数据库
    @Override
    public int update(User user) {
        return 0;
```

```java
    }

    // 查询 user 对象
    @Override
    public List<User> select(User user) {
        List<User> users = new ArrayList<User>();
        ResultSet rs = null;
        if(null == user) { // User 为空则查询全部
            String sql = "select * from t_user";
            rs = DBHelper.getResultSet(sql);
        } else { // User 不为空则查询 User
            String sql = "select * from t_user where name = ?";
            Object[] objs = {user.getName()};
            rs = DBHelper.getResultSet (sql, objs);
        }

        try {
            while(rs.next()) { // 遍历结果集
                User u = new User(rs.getInt(1), rs.getString(2), rs.getString(3),
                    rs.getString(4), rs.getString(5), rs.getString(6));
                users.add(u);
            }
        } catch (SQLException e) {
            e.printStackTrace();
        }

        return users;
    }

    // 查询 id 对应的数据库记录
    @Override
    public User selectById(String id) {
        return null;
    }

    // 查询 user 对象是否存在于数据库中
    @Override
    public boolean isExist(User user) {

    }

    // 通过用户名和密码查询 user 对象
    @Override
    public User selectByNameAndPassword(User user) throws SQLException {
        User u = null;
        String sql = "select * from t_user where name = ?and password = ?";
        Object[] objs = {user.getName(), user.getPassword()};
        ResultSet rs = DBHelper.getResultSet (sql, objs);
        if(rs.next()) { // 遍历结果集
            u = new User(rs.getInt(1), rs.getString(2), rs.getString(3),
```

```
            rs.getString(4), rs.getString(5), rs.getString(6));
    }
    return u;
  }
}
```

7. 运行项目

点击工具栏上的"Run"按钮（快捷键 Ctrl+F11），在弹出的窗口中选择"Run On Server"，启动完成后在浏览器中输入 http://localhost:8080/servlet 打开页面，效果如图 4.19 所示。

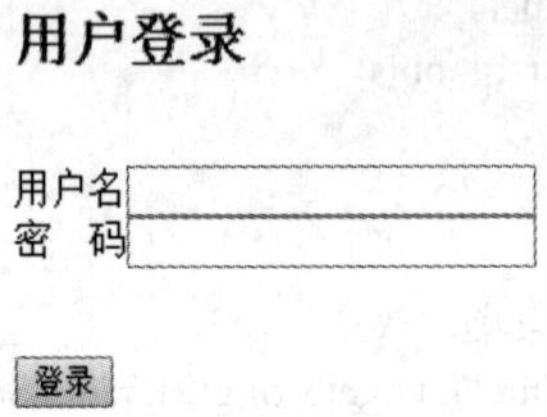

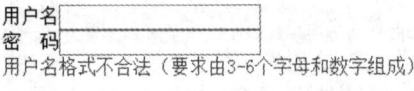

图 4.19　登录界面

如果用户填写的登录信息没有验证通过，程序会为用户响应失败信息，登录失败页面效果如图 4.20 所示。

图 4.20　登录失败页面效果

如果登录用户名和密码不匹配，程序会为用户提供登录失败信息，具体效果如图 4.21 所示。

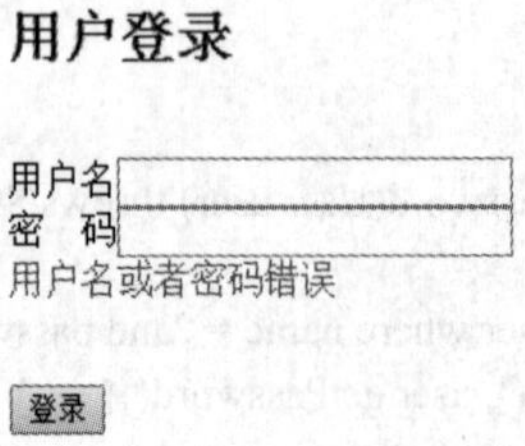

图 4.21　用户名或密码错误页面效果

如果登录成功，会显示登录成功页面，具体效果如图 4.22 所示。

欢迎：why

id	name	操作(添加)
1	王海洋	修改 删除
2	why	修改 删除
3	why2	修改 删除
4	why3	修改 删除

图 4.22　登录成功页面效果

点击"添加"链接，会进入添加界面，具体效果如图 4.23 所示。

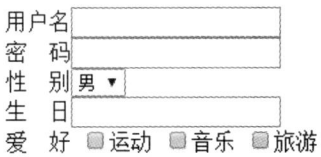

图 4.23　新增用户页面效果

如果用户填写的用户信息没有验证通过，程序会为用户响应失败信息，添加失败页面效果如图 4.24 所示。

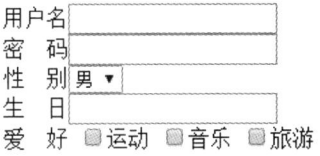

图 4.24　添加失败页面效果

添加成功后，会跳转到首页，如图 4.22 所示。

【实践训练】

继续完善任务 4.2 的功能，在管理员登录成功后，会显示用户列表，实现该用户列表中的修改和删除功能。

拓展训练

查阅相关资料，学习 JSTL 标签中其余 3 个标签库（fmt、sql 和 XML 标签库）以及 fn 方法库。

同步训练

一、填空题

1. Cookie 通过在 _____ 记录信息确定用户身份，而 Session 通过在 _____ 记录信息确定用户身份。
2. Cookie 所属类的完整名称（包名 + 类名）是 _____。
3. 实现会话跟踪的机制有 _____ 机制和 _____ 机制。
4. JSTL core 标签库中的表达式控制标签主要包括 <c:out> 标签、<c:set> 标签、<c:remove> 标签和 _____ 标签。
5. JSTL 中使用 _____ 标签实现类似于 Java 中的 while 和 for 的功能。

二、简答题

1. 简述 Session 和 Cookie 的主要区别。
2. 简述 Session 和 Cookie 的工作原理。
3. 在 JSP 中实现数据输出的方式有几种？各自的实现原理是什么？
4. 如何使用 JSTL 标签实现类似于 Java 中 if…else 的功能？

项目 5
实现网上书店

单元介绍

以网上书店应用系统为例,了解 Web 应用系统的用户角色和功能模块划分方法,掌握 Web 应用系统的需求分析、系统设计、代码编写和测试等设计流程。

本项目要实现的功能如下:
- 用户注册、登录
- 浏览、搜索图书
- 选购图书、提交订单

学习目标

【知识目标】
　　了解 Web 应用系统的需求分析和功能模块的设计方法
　　掌握分层设计方法在具体系统中的应用
　　掌握 Servlet 技术在具体系统中的应用

【能力目标】
　　能根据实际情况分析 Web 应用系统的用户需求,设计功能模块
　　能应用分层设计方法实现功能模块
　　能应用 Servlet 技术实现业务流程处理和跳转

任务 5.1 用户登录和图书展示模块设计

【任务分析】

整个系统主要包括用户注册和登录、浏览和搜索图书、选购图书和提交订单等功能。系统设计采用 MVC 架构，Servlet 充当控制（Controller）层，JSP 页面充当视图（View）层，模型（Model）层由则包括实体类和数据访问对象（DAO）。本次任务完成用户登录和图书展示模块设计。

【相关知识】

5.1.1 网上书店系统需求分析

该系统的使用者是浏览商品、购物的顾客，通过分析，用户需求如下：
①有统一、友好的操作界面，可以通过菜单导航到各个子界面。
②提供完善的编辑功能，对基础信息进行录入和修改。
③可以按类别查看图书信息。
④能根据图书名称进行模糊查询。
⑤能在线购书，生成订单。

5.1.2 功能模块设计

系统功能主要分为用户认证和在线购书两部分，其中用户认证功能包括用户注册、用户登录、修改用户信息子模块；在线购书包括分类浏览图书、查找图书、加入购物车、修改购物数量和提交订单子模块，如图 5.1 所示。

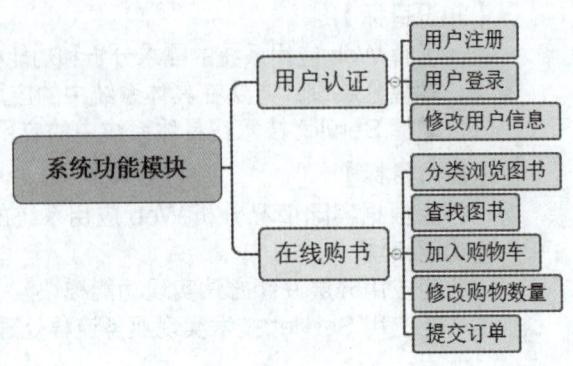

图 5.1 系统功能模块

1. 页面布局设计

网上书店的页面采用统一布局方式,整个页面分为三部分:第一部分为通用的网页头部,第二部分是各个页面的主要内容,第三部分是网页尾部,如图 5.2 所示。

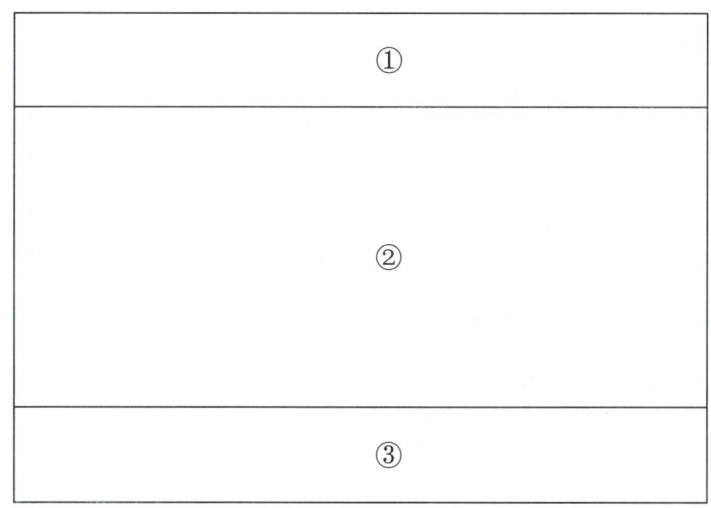

图 5.2　前台部分网页整体布局

2. 设计网页头部和尾部

为了保持网站整体风格统一,对网页头部和尾部要进行统一设计。其中网页头部页面为 header.jsp、尾部页面为 tail.jsp。

网页头部中包含了网站标题、背景图、导航栏,以及搜索栏和用户信息,效果如图 5.3 所示。

图 5.3　通用的网页头部

网页头部代码如下:

header.jsp
```
<%@ page language="java" pageEncoding="utf-8"%>
<div class="header">
  <div class="logo"> 书香小屋 </div>
  <div id="menu">
```

```html
     <ul>
       <li class="selected"><a href="index.jsp"> 首页 </a></li>
       <li><a href="page?type=0"> 图书展示 </a></li>
       <li><a href="page?type=4"> 特别推荐 </a></li>
       <li><a href="okLoggin"> 个人中心 </a></li>
       <li><a href="userLoginOut"> 注销 </a></li>
     </ul>
   </div>
</div>

<div style="margin-top:5px;padding-left:10px">
  <form action="pageSearchServlet" method="post" id="searchForm">
    <table>
      <tr>
        <td><input type="text" name="keywords" id="keywords" /></td>
        <td><input type="image" src="images/search.gif" name="submit" /></td>
        <td><a href="superSearchServlet"> 高级搜索 </a></td>
        <td width="200" align="right">
          <c:choose>
            <c:when test="${sessionScope.user!=null}">
              欢迎您， <font color="red">${sessionScope.user.name}</font>
                  <a href="/LogoutServlet"> 注销 </a>
            </c:when>
            <c:otherwise>
              您好，请 <a href="login.jsp"> 登录 </a>
            </c:otherwise>
          </c:choose>
        </td>
      </tr>
    </table>
  </form>
</div>
```

网页尾部包含网站的备案号、地址、邮编、电话等联系方式，效果如图5.4所示。

备案号：xxxxxxxx 地址：xxxxxxxxxxxxx 邮编：xxxxxx 电话：xxxxxxxxx

图5.4 通用的网页尾部

网页尾部代码如下：

```html
footer.jsp
<%@ page language="java" pageEncoding="UTF-8"%>
<div class="footer">
  <div class="right_footer">
    备案号：xxxxxxxxx
    地址：xxxxxxxxxxxxxx 邮编：xxxxxx 电话：xxxxxxxxx
  </div>
</div>
```

3. 设计用户登录界面

在图 5.3 所示的网页头部中，当用户尚未登录时，网页头部下方显示"您好，请登录"链接，点击后进入用户登录界面，如图 5.5 所示。

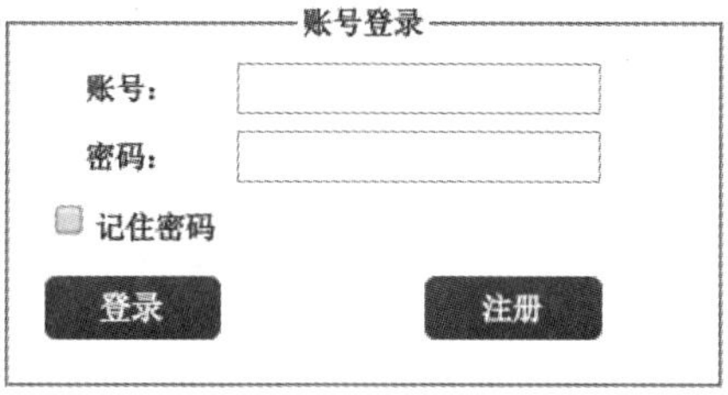

图 5.5　用户登录界面

用户登录界面代码如下：

```jsp
<%@ page language="java" import="java.util.*" pageEncoding="UTF-8"%>
<!DOCTYPE HTML PUBLIC "-//W3C//DTD HTML 4.01 Transitional//EN">
<html>
 <head>
  <title> 账号登录 </title>
 </head>
 <body>
 <div class="login">
  <form name="login" action=" LoginServlet" method="post">
  <fieldset>
     <legend id="loginTitle"> 账号登录 </legend>
     <table>
       <tr>
          <td> 账号： </td>
          <td><input type="text" name="name" /></td>
       </tr>
       <tr>
          <td> 密码： </td>
          <td><input type="password" name="password" /></td>
       </tr>
       <tr>
          <td><input type="checkbox" name="remember" /> 记住密码 </td>
          <td> </td>
       </tr>
       <tr>
          <td><input type="submit" class="register" value=" 登录 " /></td>
          <td><a class="register" href="register.jsp"> 注册 </a></td>
       </tr>
     </table>
  </fieldset>
  </form>
 </div>
 </body>
</html>
```

观察 HTML 代码可知，表单标签提交后的处理程序是 LoginServlet，提交方式为 POST，接下来要设计用户登录的后台处理模块。

4. 设计用户登录处理模块

用户登录处理模块采用分层设计方法，各层设计如下。

①数据库层

在数据库中，用 tb_user 表存储用户信息，其结构如表 5.1 所示。

表 5.1 tb_user 表结构

字段名	类型	说明
id	int	唯一主键
name	varchar(10)	用户名（不能重复）
password	varchar(10)	密码
email	varchar(30)	电子邮箱
trueName	varchar(10)	真实姓名
sex	varchar(10)	性别
birthday	date	出生日期
phone	varchar(30)	电话号码
address	varchar(100)	家庭住址

②实体类（Entity）

根据数据表结构，设计对应的实体类 User，代码如下：

```
package com.bookshop.entity;
import java.util.Date;
public class User {
    private int id;
    private String name;
    private String password;
    private String email;
    private String trueName;
    private String sex;
    private Date birthday;
    private String address;
    private String phone;
    // 字段对应的 getter 和 setter 省略（可以用 Eclipse 自动生成）
}
```

③数据访问层（DAO）

数据访问层承担直接和数据库进行交互的任务，与 User 相关的数据库操作封装在 UserDao 中，UserDao 中的 login 方法对用户名和密码进行验证，如果验证成功，返回包含用户信息的 user 对象，否则返回 null。

package com.bookshop.dao;

```java
import java.sql.*;
import com.bookshop.entity.*;
import tools.*; //tools 包中包含 DBHelper 类
public class UserDao
{
    public User login(String name,String password) {
        User user = new User();
        String sql ="select * from tb_user where name=? and password=?";
        ResultSet rs = DBHelper.getResultSet(sql,new Object[]{name,password});
        try
        {
           if(rs!=null && rs.next()) {  // 存在用户名和密码都匹配的记录
              user.setId(rs.getInt("id"));
              user.setName(rs.getString("name"));
              return user;
           }
        }
        catch(SQLException ex){ }
        return null;
    }
}
```

④控制层

控制层接收客户端提交的数据，处理后将结构存入缓存中，再跳转到指定的 JSP 页面显示结果，该层通过 Servlet 实现。

在进行数据处理时，控制层通常要先调用 Service 层的对象方法，Service 层再调用 DAO 层完成与数据库的交互。由于用户登录验证功能比较简单，故没有设计 Service 层，由控制层直接调用 DAO 层中 UserDao 类的 login 方法来完成用户信息比对。

```java
package com.bookshop.servlet;
import java.io.*;
import javax.servlet.*;
import javax.servlet.http.*;
import com.bookshop.dao.*;
import com.bookshop.entity.*;
public class LoginServlet extends HttpServlet {
    public void doPost(HttpServletRequest request, HttpServletResponse response)
    throws ServletException, IOException {
        request.setCharacterEncoding("UTF-8");
        String name = request.getParameter("name");
        String password=request.getParameter("password");
        UserDao userDao = new UserDao(); // 通过 DAO 层完成数据库操作
        User user = userDao.login(name, password);
        if(user!=null)   // 登录验证成功
           request.getSession().setAttribute("user",user); // 存储到 Session 中
        response.sendRedirect("index.jsp");
    }
}
```

LoginServlet 在 web.xml 中的配置如下：
```xml
<servlet>
    <servlet-name>LoginServlet</servlet-name>
    <servlet-class>com.bookshop.servlet.LoginServlet</servlet-class>
</servlet>
<servlet-mapping>
    <servlet-name>LoginServlet</servlet-name>
    <url-pattern>/LoginServlet</url-pattern>
</servlet-mapping>
```

⑤表现层

用户登录成功后，网页头部下方会显示欢迎字样，用户可以点击"注销"链接退出登录，如图 5.6 所示。

图 5.6　用户信息显示界面

对应的代码参考 header.jsp 中的 jstl 标签 <c:choose> 和 </c:choose> 之间部分。

5. 设计图书展示模块

图书展示页面包含选择类别、分页导航、图书列表三部分。用户可以选择图书类别，并分页浏览该类别下的图书，点击图书后显示详情，如图 5.7 所示。下面分层进行设计。

图 5.7　图书展示页面

①数据库层

在数据库中，表 tb_type 存储图书类别（结构见表 5.2），表 tb_book 存储图书详细信息（结构见表 5.3），tb_book 表的 typeId 字段引用 tb_type 表的 id 字段。

表 5.2　tb_type 表结构

字段名	类型	说明
id	int	唯一主键
typeName	varchar(20)	图书类别名（不能重复）

表 5.3　tb_book 表结构

字段名	类型	说明
id	int	唯一主键
typeId	int	图书类别 id（外键）
bookName	varchar(50)	图书名称
ISBN	varchar(20)	ISBN 编号
publisher	varchar(20)	出版社
author	varchar(20)	作者
picture	varchar(100)	对应图片文件名
introduce	text	图书介绍
price	float	价格
pages	int	页数

②实体类（Entity）

根据数据表结构，设计对应的实体类 BookType 和 Book，代码如下：

```
package com.bookshop.entity;
public class BookType {
    private int id;
    private String typeName;
    // 字段对应的 getter 和 setter 省略（可以用 Eclipse 自动生成）
}


package com.bookshop.entity;
public class Book {
    private int id;
    private int typeId;
    private String bookName;
    private String introduce;
    private float price;
    private String picture;
    private int pages;
    private String publisher;
    private String author;
    private String ISBN;
    // 字段对应的 getter 和 setter 省略（可以用 Eclipse 自动生成）
}
```

③数据访问层（DAO）

数据访问层包括 BookTypeDao 和 BookDao 两个类，其中 BookTypeDao 类的 getAllType 方法从数据表中查询所有的图书类别，并以集合形式返回，代码如下：

```java
package com.bookshop.dao;
import java.util.*;
import java.sql.*;
import com.bookshop.entity.*;
import tools.*; //tools 包中包含 DBHelper 类
public class BookTypeDao
{
    public List getAllType() { // 查询并返回所有的图书类别
        List typeList = new ArrayList();
        ResultSet rs = null;
        try {
            String sql = "select * from tb_type";
            rs = DBHelper.getResultSet(sql);
            while(rs.next()){
                BookType bt = new BookType();
                bt.setId(rs.getInt(1));
                bt.setTypeName(rs.getString(2));
                typeList.add(bt);
            }
        }
        catch (SQLException e) { }
        return typeList;
    }
}
```

BookDao 类的 getBookByType 方法通过类别 id 查询相关的图书，并且支持分页，执行后返回一个 Page 类的对象。Page 类定义如下：

```java
package tools;
import java.util.List;
public class Page {
    private int totalNum; // 总记录数
    private int currentPage; // 当前页号
    private List data; // 当前页的数据集
    // 字段对应的 getter 和 setter 省略（可以用 Eclipse 自动生成）
}
```

在执行过程中，要填写 page 对象的 3 个属性：

- 先根据图书类别 ID、当前页的起始位置 offset 和每页的记录条数 pageSize，查询出当前页的数据，存放到 page 对象的 data 属性中。
- 再根据图书类别 ID 查询该类图书的总数，存放到 page 对象的 totalNum 属性中。
- 最后根据提交的 offset 参数，计算出当前页号，存放到 page 对象的 currentPage 属性中。

```java
package com.bookshop.dao;
import java.util.*;
import java.sql.*;
```

```java
import com.bookshop.entity.*;
import tools.*; //tools 包中包含 DBHelper 类
public class BookDao
{
    public Page getBookByType(int typeId,int offset,int pageSize) {
        Page page=new Page();
        List bookList = new ArrayList();
        int totalNum=0; // 总的记录数
        ResultSet rs = null;
        try {
            String sql = "select * from tb_book where typeId=? limit "+offset+","+pageSize;
            rs = DBHelper.getResultSet(sql,new Object[]{typeId});
            while(rs.next()){
                Book book = new Book();
                book.setId(rs.getInt("id"));
                book.setBookName(rs.getString("bookName"));
                book.setAuthor(rs.getString("author"));
                book.setPicture(rs.getString("picture"));
                bookList.add(book);
            }

            Object count=DBHelper.ExecScalar("select count(*) from tb_book where typeId=?",new Object[]{typeId});
            totalNum=Integer.parseInt(count.toString());

        }
        catch (SQLException e) { }
        page.setData(bookList); //bookList 集合对应 page 对象的 data 属性
        page.setTotalNum(totalNum); // 该类别图书的总数
        page.setCurrentPage(offset/pageSize+1); // 当前页号
        return page;
    }
}
```

④控制层

图书展示的控制层由 BookShowServlet 担任，该层调用 DAO 层获得数据，并将图书类别集合（types）、当前选择的类别（selTypeId）、当前页面的相关数据（page）存放在 request 对象的键值对集合中，并推送给表现层进行显示。

```java
package com.bookshop.servlet;
import java.io.IOException;
import java.util.*;
import javax.servlet.*;
import javax.servlet.http.*;
import tools.Page;
import com.bookshop.dao.BookDao;
import com.bookshop.dao.BookTypeDao;
import com.bookshop.entity.BookType;
public class BookShowServlet extends HttpServlet {
    public void doGet(HttpServletRequest request, HttpServletResponse response)
```

```
        throws ServletException, IOException {
    doPost(request, response);
}
public void doPost(HttpServletRequest request, HttpServletResponse response)
        throws ServletException, IOException {
    // 图书类别列表
    BookTypeDao bookTypeDao=new BookTypeDao();
    List types= bookTypeDao.getAllType();
    request.setAttribute("types", types);
    // 当前选择的图书类别
    int selTypeId=0;
    try
    {
        selTypeId=Integer.parseInt(request.getParameter("typeId"));
    }
    catch(Exception ex) {
        // typeId 参数异常时,默认选择第一个图书类别 Id
        if(types.size()>0) selTypeId=((BookType)types.get(0)).getId();
    }
    request.setAttribute("selTypeId", selTypeId);
    // 当前类别、当前页包含的图书
    int offset=0;
    try{
        offset=Integer.parseInt(request.getParameter("pager.offset"));
    }
    catch(Exception ex) { }
    BookDao bookDao=new BookDao();
    Page page=bookDao.getBookByType(selTypeId,offset,10); // 每页 10 条数据
    request.setAttribute("page", page);
    request.getRequestDispatcher("bookShow.jsp").forward(request, response);
    }
}
```

⑤表现层

图书展示页面的表现层代码存放在 bookShow.jsp 中,包括图书选择类别下拉列表、分页导航和图书列表,效果如图 5.7 所示。其中选择类别下拉列表用 select 标签表示,选择之后调用 selectType 方法跳转到相应的图书类别,分页导航栏用 pager-taglib 的系列标签实现,图书列表用 JSTL 的 forEach 标签循环显示。

分页标签库 pager-taglib 系列标签的用法如下。

- pg:pager

作用:设置分页的总体参数。

格式:<pg:pager items=" 记录总数 " url=" 分页链接根目录 " maxPageItems=" 每页记录数 " maxIndexPages=" 最大页码数 ">。

- pg:first

作用:第一页标签。

可用变量：

pageUrl，分页链接 URL 地址。

pageNumber，页码。

firstItem，首页第一行的索引值。

lastItem，首页最后一行的索引值。

- pg:prev

作用：上一页标签。

可用变量：

pageUrl，分页链接 URL 地址。

pageNumber，页码。

firstItem，前页第一行的索引值。

lastItem，前页最后一行的索引值。

- pg:pages

作用：循环输出页码信息。

可用变量：

pageUrl，分页链接 URL 地址。

pageNumber，页码。

firstItem，该页第一行的索引值。

lastItem，该页最后一行的索引值。

- pg:next

作用：下一页标签。

可用变量：

pageUrl，分页链接 URL 地址。

pageNumber，页码。

firstItem，下页第一行的索引值。

lastItem，下页最后一行的索引值。

- pg:last

作用：下一页标签。

可用变量：

pageUrl，分页链接 URL 地址。

pageNumber，页码。

firstItem，尾页第一行的索引值。

lastItem，尾页最后一行的索引值。

图书展示页面的表现层代码如下：

```
<%@ page language="java" import="java.util.*" pageEncoding="UTF-8"%>
<%@taglib uri="http://java.sun.com/jsp/jstl/core" prefix="c"%>
<%@taglib uri="http://jsptags.com/tags/navigation/pager" prefix="pg"%>
<!DOCTYPE HTML PUBLIC "-//W3C//DTD HTML 4.01 Transitional//EN">
```

```html
<html>
 <head>
  <title> 账号登录 </title>
   <link rel="stylesheet" type="text/css" href="style.css" />
   <script type="text/javascript">
   function selectType() { // 下拉框选择事件处理
     var typeId = document.getElementById("type").value;
     window.location="BookShowServlet?typeId="+typeId;
   }
   </script>
 </head>
 <body>
 <div id="pageborder">
  <%@include file="header.jsp" %>
  <table>
   <tr>
     <td> 选择类别：</td>
     <td>
      <select id="type" name="type" onchange="selectType()">
        <c:forEach var="flag" items="${types}">
         <option value="${flag.id}"
           <c:if test="${flag.id==selTypeId}">selected</c:if>
           >${flag.typeName}</option>
        </c:forEach>
      </select>
     </td>
     <td><div class="pager">
       <!-- 分页标签 -->
         <pg:pager items="${page.totalNum}" url="BookShowServlet" maxPageItems="10" maxIndexPages="20">
         <pg:first><a href="${pageUrl}&typeId=${selTypeId}"> 首页 </a></pg:first>
         <pg:prev><a href="${pageUrl}&typeId=${selTypeId}"> 前页 </a></pg:prev>
         <pg:pages><a
         <c:if test="${pageNumber==page.currentPage }">
         style="text-decoration:underline"
         </c:if>
         href="${pageUrl}&typeId=${selTypeId}">${pageNumber}</a></pg:pages>
         <pg:next><a href="${pageUrl}&typeId=${selTypeId}"> 后页 </a></pg:next>
         <pg:last><a href="${pageUrl}&typeId=${selTypeId}"> 尾页 </a></pg:last>
         </pg:pager>
        </div>
     </td>
   </tr>
  </table>
  <c:forEach var="book" items="${page.data}">
    <div class="product_box">
     <a href="ShowBookByIdServlet?bookId=${book.id}">${book.bookName}</a>
      <div class=" product_bg">
       <a href="showBookByIdServlet?bookId=${book.id}"><img src="images/${book.picture}" class="thumb" border="0" /></a>
```

```
        </div>
      </div>
    </c:forEach>
</div>
</body>
</html>
```

在上面的代码中，分页链接的 URL 地址为 "${pageUrl}&typeId=${selTypeId}"，解析后的链接地址为 "BookShowServlet?pager.offset= 该页起始位置 &typeId= 当前类别 ID"，大家可以通过查看网页源码观察生成的分页链接地址规律。

同时，为了给当前页码作一个标记，用 JSTL 的 if 标签判断是否为当前页，满足条件则增加下划线样式，对应的代码是：

```
<c:if test="${pageNumber==page.currentPage}">
style="text-decoration:underline"
</c:if>
```

【实践训练】

1. 完成网上书店用户注册功能。
2. 完成图书搜索功能。

任务 5.2　购物和订单生成模块设计

【任务分析】

用户在浏览商品的过程中，可以将感兴趣的商品添加到购物车，最后通过购物车生成订单，本任务完成购物车和订单生成功能模块设计。

【相关知识】

5.2.1　网上商城购物车

在现实生活中，购物车是为了在购物过程中方便顾客设计的。有了购物车，顾客不仅可以轻松携带大量商品，还可以解放双手来选购其他商品。

在网上商城购物时，商品都是一些数据，购物车则是起到存储多条数据的作用，为合并购买提供方便。当顾客将多件商品放入购物车时，可以一次性输入收件人信息并提交订单、支付货款。此外，购物车还具有收藏功能，可以将顾客喜欢但不急需购买的商品放入购物车中收藏。购物车通常具有以下功能：

①把商品添加到购物车,即订购商品。
②删除购物车中已订购的商品。
③修改购物车中某一商品的订购数量。
④清空购物车。
⑤显示购物车中商品清单,包括数量、单价、总价。

5.2.2 购物车的数据存储方式

(1)用 Session 实现页面之间数据共享

实现购物车的关键在于服务器能识别每一个用户并保持联系,但是 HTTP 是一种无状态的协议,因而服务器不能记住是谁在购买商品,当把商品加入购物车时,服务器也不知道购物车里原先有些什么,用户在不同页面间跳转时购物车无法"随身携带"。

为了存储用户数据,服务器为每个访问网站的用户开辟了一块存储空间,称为 Session,它是用户私有的存储区域,服务器还给客户端浏览器发送了一把钥匙,称为 Session ID,只有知道 Session ID,才能取出 Session 里面存储的数据。客户端收到 Session ID 后,存放在浏览器磁盘上的特定存储区域中,该区域称为 Cookie。在不同页面之间跳转时,就可以通过 Session ID 这把钥匙来识别用户身份并取出存放在服务器端的用户私有数据。

(2)购物车的数据结构选择

购物车中顺序存放了多条商品的数据,本质是上一个线性表,但又希望能直接通过商品的 ID 找到购物车中对应的商品数量,而无需对线性表进行遍历。基于以上两点考虑,选择了 Java 中的 LinkedHashMap 类来存储购物车数据,既保证了记录的有序性,又能通过关键字快速获取到对应的值。

【任务实施】

1. 设计图书详细信息显示模块

当用户对某一本图书感兴趣时,可以点击超链接进一步查看图书详情,该功能模块分层设计如下:

①数据表和实体类

数据表使用 tb_book,实体类使用 Book 类,前面已经作了设计。

②数据访问层(DAO)

在 BookDao 类中增加 getBookById 方法,以图书 ID 作为参数查询出图书的详细信息,并返回 Book 对象。

```
package com.bookshop.dao;
import java.util.*;
import java.sql.*;
import com.bookshop.entity.*;
import tools.*; //tools 包中包含 DBHelper 类
```

```java
public class BookDao
{
    public Book getBookById(int bookId){
        Book book = new Book();
        ResultSet rs = null;
        String sql = "select * from tb_book where id=?";
        try {
            rs = DBHelper.getResultSet(sql,new Object[]{bookId});
            if(rs.next()){
                book.setId(rs.getInt("id"));
                book.setBookName(rs.getString("bookName"));
                book.setIntroduce(rs.getString("introduce"));
                book.setISBN(rs.getString("ISBN"));
                book.setAuthor(rs.getString("author"));
                book.setPages(rs.getInt("pages"));
                book.setPublisher(rs.getString("publisher"));
                book.setPrice(rs.getFloat("price"));
                book.setPicture(rs.getString("picture"));
            }
        }
        catch (SQLException e) { }
        return book;
    }
}
```

③控制层

控制层接收通过 URL 提交的 bookId 参数，通过该参数调用 DAO 层的 getBookById 方法，获取 Book 对象，存放在 Request 对象的键值对表中，再跳转到 details.jsp 进行显示。

```java
package com.bookshop.servlet;
import java.io.IOException;
import javax.servlet.ServletException;
import javax.servlet.http.HttpServlet;
import javax.servlet.http.HttpServletRequest;
import javax.servlet.http.HttpServletResponse;
import com.bookshop.dao.BookDao;
import com.bookshop.entity.Book;
public class ShowBookByIdServlet extends HttpServlet {
    public void doGet(HttpServletRequest request, HttpServletResponse response)
            throws ServletException, IOException {
        int bookId = 0;
        try
        {
            bookId = Integer.parseInt(request.getParameter("bookId"));
            //System.out.println(bookId);
        }
        catch(Exception ex) { }
        BookDao bookDao=new BookDao();
        Book book = bookDao.getBookById(bookId);
        request.setAttribute("book", book);
        request.getRequestDispatcher("details.jsp").forward(request, response);
    }
}
```

④表现层

图书详细信息展示界面如图 5.8 所示，界面中显示图书名称、ISBN 编号、页数、作者、出版社、价格、图片、简介等信息，用户可以点击"立即购买"按钮生成订单。

图 5.8 图书详细信息展示页面

图书详细信息展示 details.jsp 的代码如下：

```jsp
<%@ page language="java" import="java.util.*" pageEncoding="UTF-8"%>
<%@taglib uri="http://java.sun.com/jsp/jstl/core" prefix="c"%>
<!DOCTYPE HTML PUBLIC "-//W3C//DTD HTML 4.01 Transitional//EN">
<html>
<head>
<title> 图书详情 </title>
<link rel="stylesheet" type="text/css" href="style.css" />
</head>
<body>
<div id="pageborder">
<%@include file="header.jsp" %>
<div class="left_content">
<div class="title"><span class="title_icon"><img src="images/flower.gif" /></span>
图书详情 </div>
    <div class="product_details"> <!-- product_details 开始 -->
    <div class="product_img"><img src="images/${book.picture}" border="0" />
    </div>
    <table>
     <tr>
        <td width="100"> 书籍名称 </td>
        <td width="150">${book.bookName}</td>
        <td width="300" rowspan="6">${book.introduce}</td>
    </tr>
    <tr>
        <td>ISBN</td> <td>${book.ISBN}</td>
    </tr>
    <tr>
        <td> 页数 </td> <td>${book.pages}</td>
```

```html
      </tr>
      <tr>
        <td> 作者 </td> <td>${book.author}</td>
      </tr>
      <tr>
        <td> 出版社 </td> <td>${book.publisher}</td>
      </tr>
      <tr>
        <td> 价格 </td> <td>${book.price}</td>
      </tr>
    </table>
<a href="buyBooksServlet?bookId=${book.id}" class="more">
<img src="images/buy.jpg" border="0" /></a>
 </div><!-- product_details 结束 -->
</div>
</body>
</html>
```

2．设计购物车模块

当用户在浏览商品时，如果对某件商品感兴趣，可以临时加入到购物车中，就像在超市选购货物一样。从软件设计角度看，购物车是存在于 Session 中的对象，存储了选购商品 ID、数量等信息，即使用户在选购时尚未登录，也可以存放数据到 Session 中。

①设计购物车类

购物车类的代码存放在 Cart.java 中，包含了购物车的一些基本操作，包括添加商品、修改商品数量、获取商品数量、删除商品、统计总价格等功能。购物车本质上是一个数据集合，既要求集合元素关键字（商品 ID）不能重复，又要求商品添加的顺序能被记录，所以采用 LinkedHashMap 类来存储选购的商品，每个集合元素为一个键值对，其中关键字为商品 ID，对应值为商品数量，如表 5.4 所示。

表 5.4　购物车数据结构

商品 ID	商品数量
1	3
5	2
3	1

Cart.java 代码如下：

```java
package tools;
import java.util.*;
import com.bookshop.dao.BookDao;
import com.bookshop.entity.Book;
public class Cart {
  private LinkedHashMap<Integer, Integer> items;
  public Cart() {
    items = new LinkedHashMap<Integer, Integer>();
  }
```

```java
// 添加一件商品到购物车中
public synchronized void addItem(Integer bookId) {
    if(items.containsKey(bookId))
        items.put(bookId, items.get(bookId)+1);
    else
        items.put(bookId, 1);
}
// 修改购物车中指定商品的数量
public synchronized void setItemCount(Integer bookId,Integer count) {
    items.put(bookId, count);
}
// 删除购物车中的商品
public synchronized void deleteItem(Integer bookId){
    if(items.containsKey(bookId))
        items.remove(bookId);
}
// 获取购物车中指定商品的数量
public synchronized Integer getBookCount(Integer bookId){
    return items.get(bookId);
}
// 清空购物车
public synchronized void clear(){
    items.clear();
}
// 计算购物车中商品的总价格
public synchronized float getTotalPrice(){
    float total = 0.0f;
    BookDao bookDao=new BookDao();
    Iterator<Integer> it = items.keySet().iterator();
    while(it.hasNext()){
        Integer key = it.next();
        total+=bookDao.getBookPriceById(key)*items.get(key);
    }
    return total;
}
// 获取购物车中的商品列表
public synchronized List getSelBooks()
{
    List books=new ArrayList();
    BookDao bookDao=new BookDao();
    Iterator<Integer> it = items.keySet().iterator();
    while(it.hasNext()){
        Integer key = it.next();
        Book book=bookDao.getBookById(key);
        books.add(book);
    }
    return books;
}
}
```

②添加商品到购物车

在图 5.8 中，用户查看图书信息后，点击"立即购买"按钮，可以将该图书添加到购物车，如果曾经添加过，则购物车中该图书的数量加 1，这项功能在 BuyBooksServlet.java 中实现，代码如下：

```java
package com.bookshop.servlet;
import java.io.IOException;
import java.io.PrintWriter;
import java.util.List;
import javax.servlet.ServletException;
import javax.servlet.http.HttpServlet;
import javax.servlet.http.HttpServletRequest;
import javax.servlet.http.HttpServletResponse;
import tools.Cart;
import com.bookshop.dao.BookDao;
import com.bookshop.entity.Book;
public class BuyBooksServlet extends HttpServlet {
   public void doGet(HttpServletRequest request, HttpServletResponse response)
       throws ServletException, IOException {
     int bookId = 0;
     try
     {
       bookId = Integer.parseInt(request.getParameter("bookId"));
     }
     catch(Exception ex) { }
     BookDao bookDao=new BookDao();
     Book book = bookDao.getBookById(bookId);
     Cart cart = (Cart)request.getSession().getAttribute("cart");
     if(cart == null){ //session 里面还没有 Cart 对象
       cart = new Cart();
       request.getSession().setAttribute("cart", cart);
     }
     if(bookDao.hasBook(bookId))    cart.addItem(bookId);
     response.sendRedirect("ShowCartServlet");
   }
}
```

③显示购物车中的商品

添加商品完毕时，需要向用户展示购物车中包含商品的详细情况，并且在任何时候，用户都可以查看购物车中有哪些商品。因此，专门设计了购物车展示页面。

购物车展示页面通过控制层的 ShowCartServlet 获取数据，再转到表现层的 cart.jsp 显示结果。ShowCartServlet.java 代码如下：

```java
package com.bookshop.servlet;
import java.io.IOException;
import java.io.PrintWriter;
import java.util.List;
import javax.servlet.ServletException;
import javax.servlet.http.HttpServlet;
import javax.servlet.http.HttpServletRequest;
```

```java
import javax.servlet.http.HttpServletResponse;
import tools.Cart;
public class ShowCartServlet extends HttpServlet {
    public void doGet(HttpServletRequest request, HttpServletResponse response) throws ServletException, IOException {
        List selBooks=null;
        Cart cart=(Cart)request.getSession().getAttribute("cart");
        if(cart!=null)
           selBooks = cart.getSelBooks();
        request.setAttribute("selBooks", selBooks);
        request.setAttribute("totalPrice", cart.getTotalPrice());
        if(request.getParameter("action")==null)
           request.getRequestDispatcher("cart.jsp").forward(request, response);
        else if(request.getParameter("action").equals("order"))
        {
           if(request.getSession().getAttribute("user")==null)
              response.sendRedirect("login.jsp");
           else
              request.getRequestDispatcher("order.jsp").forward(request, response);
        }
    }
}
```

ShowCartServlet 获取购物车中的图书商品集合及其总价格，存放到 Request 对象中，再转到 cart.jsp 中进行显示。表现层 cart.jsp 的代码如下：

```jsp
<%@ page language="java" import="java.util.*" pageEncoding="UTF-8"%>
<%@taglib uri="http://java.sun.com/jsp/jstl/core" prefix="c"%>
<!DOCTYPE HTML PUBLIC "-//W3C//DTD HTML 4.01 Transitional//EN">
<html>
<head>
<meta http-equiv="Content-Type" content="text/html; charset=utf-8" />
<title> 我的购物车 </title>
<link rel="stylesheet" type="text/css" href="style.css" />
</head>
<body>
<div id="pageborder">
   <%@include file="header.jsp" %>
     <div class="left_content">
       <div class="title"><span class="title_icon">
<img src="images/flower.gif"/></span> 购物车 </div>
       <div>
         <table class="cart_table">
           <tr class="cart_title">
              <td> 图片 </td><td> 书籍名 </td><td> 单价 </td><td> 数量 </td>
              <td> 总价 </td><td> </td><td> </td>
           </tr>
           <c:forEach var="book" items="${selBooks}">
           <form action="UpdateBookCount?bookId=${book.id}" method="post">
           <tr>
              <td><img src="images/${book.picture}" width="70" height="70"/></td>
```

```html
                <td>${book.bookName}</td>
                <td>${book.price}</td>
                <td><input type="text" name="count" value="${sessionScope.cart.getBookCount(book.id)}" size="4" /></td>
                <td>${book.price*sessionScope.cart.getBookCount(book.id)}</td>
                <td><a href="DeleteItemServlet?bookId=${book.id}"> 删除 </a></td>
                <td><input type="submit" value=" 修改 "></td>
            </tr>
            </form>
        </c:forEach>
        <tr>
            <td colspan="4" class="cart_total"> 总价 :${totalPrice}</td>
            <td></td>
        </tr>
    </table>
    <a href="BookShowServlet" class="button"> 继续购物 </a>
    <a href="ShowCartServlet?action=order" class="button"> 结算 &gt;&gt;</a>
    </div>
  </div><!--end of left_content-->
</div>
</body>
</html>
```

购物车显示页面如图 5.9 所示，用户可以查看、修改、删除购物车中的项目，可以点击"继续购物"回到图书展示界面，或点击"结算"进入订单提交界面。

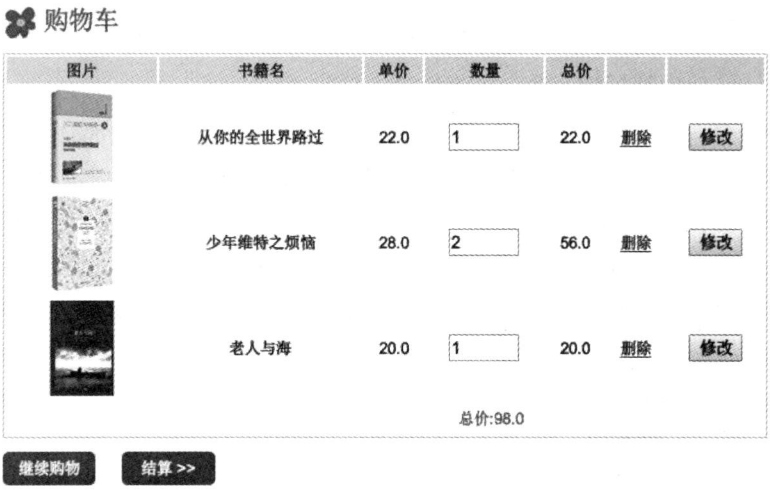

图 5.9 购物车显示页面

修改购物车中的商品数量功能对应的代码为 UpdateBookCount.java，修改完毕后，再转到 ShowCartServlet 显示新的数量。

```java
package com.bookshop.servlet;
import java.io.IOException;
import java.io.PrintWriter;
```

```java
import javax.servlet.ServletException;
import javax.servlet.http.HttpServlet;
import javax.servlet.http.HttpServletRequest;
import javax.servlet.http.HttpServletResponse;
import tools.Cart;
public class UpdateBookCount extends HttpServlet {
    public void doGet(HttpServletRequest request, HttpServletResponse response)
        throws ServletException, IOException {
      doPost(request, response);
    }
    public void doPost(HttpServletRequest request, HttpServletResponse response)
        throws ServletException, IOException {
      try
      {
        int bookId = 0;
        int count = 0;
        bookId = Integer.parseInt(request.getParameter("bookId"));
        count = Integer.parseInt(request.getParameter("count"));
        Cart cart = (Cart)request.getSession().getAttribute("cart");
        cart.setItemCount(bookId, count);
      }
      catch(Exception ex){ }
      response.sendRedirect("ShowCartServlet");
    }
}
```

删除购物车中商品调用 DeleteItemServlet 完成，对应代码是 DeleteItemServlet.java，删除后仍然转到 ShowCartServlet 中刷新显示。该功能比较简单，请读者自己完成。

3. 设计订单提交模块

在图 5.9 所示的购物车界面中点击"结算"，进入图 5.10 所示的订单填写界面，在这里填写收件人姓名、地址、邮编、手机号等信息，用于邮寄商品。

"结算"链接对应的 URL 为 ShowCartServlet?action=order，从 ShowCartServlet.java 的代码中可以看出，当请求参数 action 等于 order 时，判断用户是否登录，若用户已经登录，转到订单提交界面 order.jsp，若未登录，转到用户登录界面 login.jsp。

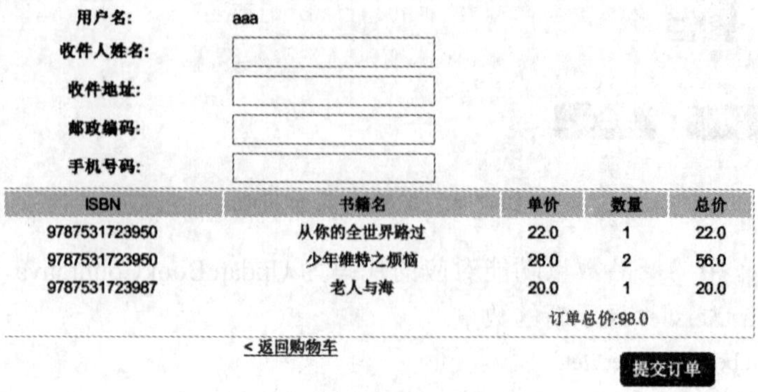

图 5.10 订单填写界面

订单提交界面 order.jsp 代码如下（客户端验证的 JavaScript 代码略）：

```jsp
<%@ page language="java" import="java.util.*" pageEncoding="UTF-8"%>
<%@taglib uri="http://java.sun.com/jsp/jstl/core" prefix="c"%>
<!DOCTYPE HTML PUBLIC "-//W3C//DTD HTML 4.01 Transitional//EN">
<html>
<head>
<meta http-equiv="Content-Type" content="text/html; charset=utf-8" />
<title> 填写收件人资料 </title>
<link rel="stylesheet" type="text/css" href="style.css" />
<script type="text/javascript" src="check.js"></script>
</head>
<body>
<div id="pageborder">
  <%@include file="header.jsp" %>
  <div class="left_content">
     <form name="submitOrder" action="AddOrderServlet" method="post">
       <table>
         <tr>
            <td class="label"><strong> 用户名 :</strong></td>
            <td align="left">${user.name}</td>
            <td></td>
         </tr>
         <tr>
            <td class="label"><strong> 收件人姓名 :</strong></td>
            <td><input type="text" id="recvName" name="recvName" class="contact_input" onblur="checkName()"/></td>
            <td class="remind" id="remindName"> </td>
         </tr>
         <tr>
            <td class="label"><strong> 收件地址 :</strong></td>
            <td><input type="text" id="address" name="address" class="contact_input" onblur="checkAddr()" /></td>
            <td class="remind" id="remindAddr"> </td>
         </tr>
         <tr>
            <td class="label"><strong> 邮政编码 :</strong></td>
            <td><input type="text" id="postcode" name="postcode" class="contact_input" onblur="checkPostalCode()" /></td>
            <td class="remind" id="remindPostCode"> </td>
         <tr>
            <td class="label"><strong> 手机号码 :</strong></td>
            <td><input type="text" id="mphone" name="mphone" class="contact_input" onblur="checkMPhone()" /></td>
            <td class="remind" id="checkMphone"> </td>
         </tr>
       </table>
     <div>
     <table class="cart_table">
        <tr class="cart_title">
```

```html
          <td>ISBN</td><td> 书籍名 </td><td> 单价 </td>
          <td> 数量 </td><td> 总价 </td>
      </tr>
      <c:forEach var="book" items="${selBooks}">
        <tr>
          <td>${book.ISBN}</td>
          <td>${book.bookName}</td>
          <td>${book.price}</td>
          <td>${sessionScope.cart.getBookCount(book.id)}</td>
          <td>${book.price*sessionScope.cart.getBookCount(book.id)}</td>
        </tr>
      </c:forEach>
      <tr>
        <td colspan="4" class="cart_total">
订单总价 :${sessionScope.cart.getTotalPrice()}</td>
      </tr>
    </table>
    <a href="ShowCartServlet" class="continue">&lt; 返回购物车 </a>
        <span class="submit_order"><input type="button" id="subOrder" name="subOrder" class="register" value=" 提交订单 " onclick="checkAll()" /></span>
      </div>
    </form>
     </div><!--end of left_content-->
</div>
</body>
</html>
```

【实践训练】

1. 完成订单提交后验证、保存功能。
2. 完成"个人中心"设计,能在里面查看购物车、管理订单。

拓展训练

1. 查阅资料,编写订单提交界面中客户端验证的 JavaScript 代码。
2. 查阅资料,在网页侧边栏中显示购物车,侧边栏平时停靠在网页右边,鼠标点击时弹出。

同步训练

一、填空题

1. MVC 是 _____、_____、_____ 的缩写。

2．在 MVC 架构中，Servlet 充当 _____ 层，JSP 充当 _____ 层。

3．JSTL 中的分支语句可以使用 _____ 标签和 _____ 标签。

4．案例中的购物车内容存放在 _____ 中，可以在多个页面之间共享。

二、简答题

1．在图书展示模块中，MVC 这三部分分别对应哪些程序？

2．购物车中的商品列表为什么选用 LinkedHashMap 来存储？

3．从控制层跳转到表现层可以用什么语句？什么情况下浏览器地址栏内容会发生改变，什么情况下不会？

参考文献

[1] 马蕾,赵婉芳. Web前端技术:XHTML+CSS+JavaScript[M]. 北京:北京邮电大学出版社,2013.

[2] 谢先伟,梅青平. Java程序设计[M]. 北京:中国水利水电出版社,2016.

[3] 刘京华. Java Web整合开发王者归来[M]. 北京:清华大学出版社,2010.

[4] 牛德雄,陈华政,李彬,扶卿妮. 基于MVC的JSP软件开发案例教程[M]. 北京:清华大学出版社,2014.

[5] 耿祥义,张跃平. JSP实用教程(第三版)[M]. 北京:清华大学出版社,2015.

[6] 郑阿奇. JSP编程教程[M]. 北京:电子工业出版社,2012.

[7] 张志锋. JSP程序设计与项目实训教程[M]. 北京:清华大学出版社,2012.

[8] 郑阿奇. Java EE基础实用教程(第2版)[M]. 北京:电子工业出版社,2014.

[9] 王斐. Java Web开发基础——从Servlet到JSP[M]. 北京:清华大学出版社,2014.

[10] 颜志军. JSP与Servlet程序设计实践教程[M]. 北京:清华大学出版社,2012.